Four Angels at Your Bed

An Inspirational Memoir of Three Generations of Women

by Carol Walker Bouldin

~ RIVERSIDE BOOKS ~

To order, contact:

SunRise Publications
P.O. Box 498763
Cincinnati, OH 45249-8763

ISBN: 0-9677569-0-1
Library of Congress Catalog Card Number: 99-97681

Design, production, and editing
by Lynn Robbins, Riverside Books,
P.O. Box 64, Camp Dennison, OH 45111

Author photo by Laura Hannam, Portland, Oregon

Printed in the USA by

MORRIS PUBLISHING

3212 East Highway 30 • Kearney, NE 68847 • 1-800-650-7888

In loving memory
of Momma, Aunt Minnie, and Daddy

And to
Kimberly Ann, my daughter,
Kristen Nicole, my niece,
Kevin Lionel & Michael Albert, my nephews, and
Brandon Luke & Devon Christopher, my grandsons:

May each of you share your Wisdom with your children and instill in them the principles of Faith in God, Love of Family, and Respect for Others as well as for Yourself.
In the words of my father, your children should be taught:
"If at once you don't succeed, try, try again!"

acknowledgments

Without the love, support, and encouragment of my husband Eugene, my father Albert, my daughter Kimberly, and my sister Marsha, the successful completion of this labor of love would not have been possible. To each of them I owe sincere thanks.

Special thanks also go to Lynn Robbins. Lynn went beyond her role as editor to provide me with the encouragement, advice, and support which allowed this book to become a reality. Her willingness to give of her time, listen to my stories, and help me unravel my thoughts are greatly appreciated. The path that led me to Lynn was a Blessing, and at the end of that path, I found not just an editor but a friend.

to family and friends...

As a member of my family circle, you may find this story difficult to read, but it is my sincere wish that you will take time to process the legacy of Momma and Aunt Minnie, two extraordinary women, and that you will value this book and pass it along to others you hold dear.

I would ask you to remember that the only way I could tell this story was from my own perspective, based on my own feelings, experiences, and perceptions. Also, because the story is told as I passed through the events, I was unable to capture every detail or include every person who provided support and caring. Please forgive my omissions and know that I am grateful to you all.

May the Lord's Blessing be with us always!

one

It was December, early evening, when the phone rang and woke me from a nap. I knew something was wrong, because it was Daddy, and Daddy never called anybody. He said Aunt Minnie had been taken to the hospital by the life squad. She was found on the floor in her apartment. Momma had already left for the hospital.

As I put down the phone, I was scared. I sat on the side of the bed, my mind racing. I had talked to Aunt Minnie just a couple of days ago, and she seemed her usual, crazy self. I wondered if I should go to the hospital. I tried to convince myself it wasn't that bad. I was tired and wanted to go back to sleep.

Suddenly I thought about Momma. Everything that happened in our family always fell on Momma's shoulders. She never complained, just went about doing whatever needed to be done. Her way was always very quiet and gentle, but her resolve was that of a good Samaritan. That's why we called her the Florence Nightingale of the family, and not just for our immediate family, but for anyone in need. So I was sure Momma could take care of things with Aunt Minnie. However, I felt I needed to go to the hospital, for Momma. And although Daddy didn't say so, I'm certain he expected me to go stay with her.

It was already dark as I headed down the expressway. I couldn't believe something was wrong with my Aunt Minnie. Daddy must not have the story straight. But then I hadn't talked to her in a couple days, and we usually talked every evening. She would call and ask Gene, "Where's badness?" Then I would have to hear about her most recent letter from the Sweepstakes. She received one or two a day, each saying she was a finalist. There was no telling Aunt Minnie that she wasn't going to win. We would laugh and talk about the money. She always assured me that she would take care of me. Since some of the letters described coming to New York, staying in a fabulous hotel suite, and touring the city, Aunt Minnie would say that I'd have to represent her, that she just wanted the money. She wanted no part of airplanes or traveling.

As I arrived at the hospital, I turned into the emergency parking lot. Driving past the emergency room window, I saw a small, frail woman with a distinct black hat. Of course, that was my Momma. When I walked in the door and started toward her, she looked up at me. There was that big smile, Momma saying, "Oh, you didn't have to come all the way over here."

"Momma what's happening?"

"I was told to wait out here. They said the apartment manager found Minnie on the floor in her bedroom, under the table, delirious. She hadn't been seen in the building for three days. She hadn't gotten her mail. She was evidently pretty disoriented."

"Momma, I'm sure I talked to her two days ago. She was fine. At least I had no reason to think anything was wrong. I'm going to find out what's going on."

This is my Momma. She would wait patiently until someone decided to give her information. Of course, I, Carol, the impatient one, had to have answers immediately. I went to the nursing station and said I wanted to know the condition of my aunt and that I wanted to talk to the doctor. The woman was very kind. She said Aunt Minnie was stable and that we could go back to see her. She would also make the doctor aware that her family was there.

As we entered the emergency room cubicle, here lies Aunt Minnie in her slip, looking all weird.

"Harriet, Carol Ann, you didn't have to come over here. I'm OK. I'm going home."

"Hold your horses," I said. "We have to talk to the doctor first, and he will decide if you're well enough to go home."

Laughing and cutting up with one of her jokes, she insisted she was fine. I told her that the emergency room doctor would have to call *her* doctor, that that was how the decision would be made. This seemed to satisfy her for the moment.

After a while, I talked Momma into going home. It was close to eight o'clock, and I knew she was tired. I went outside and watched as Momma walked across the street to the garage. I smoked two cigarettes waiting for her car to come out and head up the street. I knew that Daddy was home worrying. He would never admit this. His whole personality and temperament changed after he lost sight in his left eye four years before. Bitterness often consumed him because he could no longer do his oil painting or sculpting or write his short stories, the things he enjoyed most. However, even though he always said Minnie had no sense, this was his last surviving sibling. And despite his sharp tongue and critical nature, Daddy's heart is truly gigantic.

Suddenly, as I stood there smoking, I realized it was

freezing cold. I began to think there was no way Aunt Minnie should go home. I quickly went back inside and found the nurse. They were still waiting for her doctor to call. They felt she had mild dementia and possibly some type of infection. I began my pressure to insist they keep her for observation. She was no longer able to care for herself.

I knew I couldn't get her to come home with me, either. After all, this was a woman who wouldn't leave her apartment building. Other than going to the lobby for her mail, visiting people in the building, and occasionally going with Momma to the doctor, she would not go out. Anywhere! She even ordered her groceries and had them delivered. It didn't make any difference to her whether it was Thanksgiving, Christmas, or whatever, she had an excuse. She was as close to being a recluse as you can get. Still the word recluse doesn't exactly fit, because she truly loved people and kept up with everything going on in the apartment building and at church. Strange, but there was something about her actually going out of that building. None of us ever understood this part of her personality.

All this was going through my mind as I sat in the cubicle with Aunt Minnie drifting in and out of sleep. She continued to remind me that she was fine and wanted to go home, insisting she was OK, but I wasn't so sure. She couldn't seem to tell me what had happened. The apartment manager broke the chain on her door to get into the apartment. She was on the floor. Her story really didn't make sense. She claimed she had crawled to the refrigerator to get something to eat. She never admitted that she was found under the table in the bedroom, curled up in a fetal position, soiled. I asked why she didn't call me, and she cracked one of her jokes. She said she had been on the floor for three days. I still couldn't figure out how that was true.

As midnight approached, I became more impatient. I was glad I had persuaded Momma to go home. I talked to the doctor again, and they were still deciding whether to keep her. I was getting ticked. I thought it was probably an insurance issue. When they indicated that her tests showed she was dehydrated, possibly malnourished, and possibly had a kidney infection, I became even more insistent that she stay. Finally Aunt Minnie's doctor called. Thank goodness they would admit her.

I told Aunt Minnie she was being admitted and that her doctor would be in later in the morning. Since she loves attention, she didn't object. She seemed to fully understand what was going on and wanted me to go home. I tried to reassure her, telling her they would take her up to her room shortly and then she could get some rest. I would be back the next day, but I had to go to work first. She understood.

On the way home, I realized how tense I was. This night didn't seem real. I was sure things were going to be alright, but I had to go to work in the morning, then get myself back over to the hospital, and I was tired. I was so very tense and tired. When I got home, I filled my husband, Gene, in on what was happening. Aunt Minnie was crazy about Gene. She liked to kid and joke with him. He gave her plenty of attention, so that made him special in her book. And Gene thought Aunt Minnie was a riot, so their feelings were mutual.

Hearing of her condition, Gene seemed to feel the same as I did, saying, "Not your Aunt Minnie. She can't be sick. She'll be fine."

two

As the days passed, my daily schedule became even more hectic. Momma and I had to visit Aunt Minnie daily and convince her that she needed to stay until the doctor finished her tests. One moment she was crying, and the next she was cracking jokes. It was decided that she needed to be transferred to the Senior Care Floor, where she could stay for 21 days. This was great because she would be getting a complete medical evaluation. I talked to the social worker, and they were able to get her daily O.T. and P.T. services. The doctors seemed to think she might have had a mild stroke. Her gait was very unsteady. She also had a kidney infection that they were able to treat with antibiotics.

Now Aunt Minnie is really a character. Although she worked her entire career in a hospital as a Nurse Aide in surgery, I'm almost certain she had never been a patient. This jolly woman of 77 years began to work her magic with the nursing staff, visitors, and patients. She paddled around the halls, gathering whatever news she could, joking and teasing everyone. She knew which patients had visitors and learned as much of their business as she could. That gave her plenty to talk about when we came to visit. When I would come in, she wanted a detailed report as to where I had been. She would tell everyone, "This is my niece, Carol Ann. She used to work for the Board of Education. Now she's retired and works for the County." This went on every day.

At least every other day, I would call the social worker or talk to the doctor. I was concerned about them sending her home. She seemed much better to me, but Momma and I knew she could no longer stay alone. Of course, she thought she was going home, and that she was just fine.

One of the nice things about the Senior Care Floor was that most of the patients had to be dressed and come to the recreation room for their meals. The room was very nicely appointed and had a large screen TV. There were activities scheduled every day. The staff learned quickly that they had to insist that Aunt Minnie participate. Once they got her to take part, she took over and had everyone in stitches. She always knew everything that was going on as far as current events and could hold a conversation or offer her opinion on almost any subject. Her knowledge base came from watching television all day and half the night, as well as reading every magazine and book you could buy from Publisher's Clearing House. Still, something was wrong. Something was not right with her. I cannot explain.

Christmas for our family was truly a tradition. My sister Marsha, her children and grandson would come home from Maryland. Of course, our daughter Kim never missed anything. She would be home from St. Louis for the holiday week.

We started the day with Christmas breakfast at Momma's, and then opened presents. Either we went to Aunt Willa's (Momma's sister) or Momma had dinner. They would switch off each year. Of course, Aunt Minnie would never come. She always had an excuse. This year she would still be in the hospital. It really didn't seem to bother her because she was getting all this attention. The recreation room was decorated beautifully and since she loved flowers, she enjoyed the poinsettia.

One day, as I came down the hall, she was in the recreation room. I went in, and as usual, she broke out with that big smile. She said, "Lord child, it's too bad outside for you to be over here! ... Did you bring me anything?"

"No, not today. I've been at work. I know you remember that I have to go to work." There was another big devilish smile. "You also know that I have to go by and check on Momma and Daddy."

Another devilish smile, then she flips back with one of her witty cracks, something like, "That's fine, but *I'm* the one in the hospital."

"Aunt Minnie, what am I going to do with you?"

Now you must understand, Aunt Minnie was spoiled rotten. I guess I have to take some of the blame for this. She and her late husband, Uncle Billy didn't have children. So when she wanted something, she got it. Not only did she get it, she had to have the best. This was the way it was. So when she asked me to bring her something, she always told me what brand. She believed in brand names and would quickly say she wouldn't use "this mess" if it wasn't what she wanted. That would make Momma really ticked, because Momma was the queen of bargain shoppers.

And catalogs. Well, Aunt Minnie got every classy catalog that's printed. Not only did she get them, but she bought all kinds of clothes, china, crystal, silver, figurines, and linens from them. That's why she no longer had any money. She was living off Uncle Billy's social security and post office pension, and her pension from the Sisters of Charity was a pitiful $44 a month. Whenever she bounced a check, she would call Albert (Daddy) since his name was on her account. Daddy and Momma would have to go over to the bank and straighten out her account. The manager was kind enough to usually waive the fees. This would bring on one of Daddy's lectures about money. Aunt Minnie could care less, because she knew she could depend on Albert and Harriet.

Christmas Day, Daddy, Kim, Marsha, and I went over to the hospital. I had bought her several pairs of slacks, tops, socks, and a sweater. I wrapped them as outfits and put each in a Neiman Marcus box. This lady never wore slacks or casual clothing. All of her clothes were ultra suede, silk blouses, wool suits, dressy dresses, and none of her things were appropriate for the hospital. She now needed things that were comfortable, easy to get on, and clothes that could be laundered.

In her apartment, she wore wool skirts and always with a silk blouse. Then she would put on one of the aprons Momma had made, with big pockets in the front. Sometimes she wore a full apron and sometimes a half apron. Every year she would buy three or four from Momma when she made them for the church Christmas Bazaar.

Marsha, Kim, and I held our breath as she opened her gifts. We knew if she didn't like them we could forget it. To our surprise, Aunt Minnie was thrilled. She began to cry and got very emotional. She even said, "This is just what I needed." It was a very touching time for all of us. The truth was that her *family* made her happy. It was not the material things.

three

The day after Christmas is Momma's big day. Every year the family comes to Momma's from all around—her two sisters and their children from Columbus, cousins from Louisville, another sister and her family from Centerville, plus all the families in town stop by. This is Harriet's and Walker's Open House. (Lots of folks call my Dad "Walker.") This is a great holiday family gathering.

The out of town folks arrive around noon, and Momma serves them a hot meal. Then in the afternoon, the local family comes by the house. The table is full of Momma's cookies, brownies, fancy desserts, eggnog, and punch. Not only does Momma make a beautiful Christmas centerpiece for the table, she has Christmas arrangements all over the house, even little bouquets in the bathrooms. Every year Daddy fusses about Momma doing all this work, but without admitting it, he enjoys the day and having everyone gathered at their home.

Earlier that day, I had gone by the hospital, so when I got to Momma's, everything was underway. Marsha was busy helping set things up, and family was all over the house. For some reason, Momma didn't have as much dessert as usual. She had not done all her usual baking, saying she was tired. I decided to go to the store and buy some things before more people came, telling Marsha not to say anything to Momma. I just wanted to make sure we'd have enough food.

When I returned from the store, a white cadillac was parked in front of the house, and I couldn't figure whose it was. As I walked in, Gene took the sheet cake from me, saying, "You'll never guess who's here!" Just then I saw Kim's friend, Toren. He had flown to Columbus, rented a car, and driven to Cincinnati to surprise her. I was tickled and couldn't believe the surprise. Of course, the cousins started teasing, everyone saying that this must be serious. I thought it was great he knew how Kim felt about her family, especially her grandparents.

My usual job was kitchen helper. I really don't mind, because in our family all the juicy gossip or family happenings occurred in the kitchen. So, while I worked, I could actively participate and contribute my two cents worth.

I was exhausted when we got home. Gene and I went to bed and left Kim and Toren watching television. I told him to make himself at home because I don't believe in company. I put fresh towels in his room, but that was it.

The next morning when I staggered out to the kitchen, Toren had already made a pot of coffee. That made for extra brownie points with me. Kim said, "Did you see my presents? Come here Daddy and see what Toren gave me."

She pulled out a large history book. They had seen it in a bookstore, and Kim had said how much she wanted it. She seemed thrilled, so Gene and I kindly acknowledged that this was nice. Then she pulled out a pen from the Warner Brothers Store. Next came a large box full of colored tissue, and in it was a huge plastic engagement ring. I cracked up, but Gene stood there silent. He said nothing. I mean he said nothing.

Suddenly, Kim pulled her left hand out of her robe pocket, and on her finger was a beautiful emerald-cut diamond ring. Gene and I both gasped and congratulated and hugged them. Gene sighed and said, "I didn't think that plastic ring was a damn bit funny." We really laughed then. It takes a lot to surprise Kim, and Toren had pulled off a winner. Later, when we all were talking, Gene was still going on about being ticked off at seeing that plastic ring.

Aunt Minnie flipped when she heard about Kim's engagement and saw her ring. Of course, she had to inspect it and give it her seal of approval. She told everyone on the floor and couldn't have been more excited if Kim were her own daughter. If she had been home, she probably would have ordered *Bride* magazine from Publisher's Clearing House.

By the end of December, the hospital staff called a care conference. Daddy, Momma, Aunt Minnie, and I met with the charge nurse, occupational therapist, physical therapist, social worker, and speech therapist. Their assessments indicated that Aunt Minnie needed 24-hour care. She could no longer live alone. This was like a bolt of lightning hitting Aunt Minnie. She didn't want to admit that what they were saying was true. For once there was no joking or wise cracks, just silence and tears. As Aunt Minnie dabbed at her eyes, the social worker talked to her in a compassionate way and told her she would work with me to help find a nursing home. My dear Aunt sat there like a little girl, tears flowing, trying to explain that she could take care of herself. For a woman with so much spunk, she was pitiful, grasping for a way to keep this from happening. Yet, I believe she knew deep down inside that this was for her own good. Still, how could she admit it? How could anyone?

My insides were churning and my eyes full of tears as I

began to talk to her. I told her I'd make sure I found a nice place for her that was close to my house. I told her that Momma had to take care of Daddy and that I would take care of her. That's why she needed to be close to my house. I believe that, because she trusted me, she gave her silent, if unwilling, consent.

By the time we left the hospital, Aunt Minnie was cracking jokes and back to her witty self. At least that's what she wanted us to believe. It was as if we had not gone through this torture of a meeting. As usual, when Aunt Minnie got silly, Daddy was agitated. He just could not understand that this was her way of handling difficult situations. As always, he said Minnie didn't have any sense, though he didn't say this to her face. He at least waited until we were outside the hospital.

Things moved into high gear. I had to visit nursing homes, go to the Human Services Department and jump through hoops to get her declared eligible for Medicaid. As I sat in that crowded waiting room one morning, I felt so bad for the people who didn't have anyone to negotiate this bureaucracy for them. Everyone deserves to be treated with dignity and respect, yet appointment times didn't mean anything. We were packed in the room like cattle waiting to be called to slaughter, and we just had to wait for our number to be called.

I finally settled on a nursing home that was about fifteen minutes from my house. I talked it up to Aunt Minnie and told her about the other places I had visited and why I chose this one. Then, all of a sudden, Aunt Minnie wanted to know about her apartment and her things. I told her that Daddy and Momma had paid the rent for January, so we would have to get everything cleared out before the first of February. My heart was racing, and there was dead silence. She sat staring at me and then finally said, "What about my things?"

I told her we'd have to move some of her things to Momma's and store her furniture at my house. After another pause which seemed to last forever, she said, "I want Kim to have my furniture." We agreed that since I had room for it, she could decide later, because when she got better she would need it again.

four

Driving home from the hospital one nasty cold night, I put in a tape and headed for the expressway singing "His Eye is on The Sparrow." I sing this song whenever I'm in need of being uplifted. It brings me back into focus. On this particular evening, my mind drifted back to when Uncle Billy died, almost twenty years before.

Momma and Daddy were on vacation, in the Orient. Aunt Minnie called me, crying, in the middle of the night, saying that Billy had a heart attack and was taken to the hospital. Within minutes, she called back and said Uncle Billy was dead. She wanted Albert, my Dad, but they weren't due back until Monday and this was early Saturday. She insisted I not come over then, so I assured her I'd be over later in the morning. I didn't know what to do for her, so I took my cousin Penny with me. Two heads are always better than one.

She and Uncle Billy had lived in this apartment for probably twenty years, with their Siamese cat, Sammy. I guess Sammy was like their child, because every year you could count on the most exquisite Christmas card from "Aunt Minnie, Uncle Billy, and Sammy."

This apartment was unbelievable. The ceiling plaster was cracked and water-stained, and the wallpaper was probably the same as when they moved in. Yet, her furnishings were all covered with crisp, white linen sheets. Aunt Minnie had beautiful inlaid mahogany tables and ornate lamps with silk shades, and everything was spotlessly clean. She was meticulous about her things. As we talked, I kept thinking, Why do they live here? They obviously didn't insist that the landlord do anything to the apartment. It was weird.

Penny and I didn't know how to act. One minute Aunt Minnie was crying and sobbing, "Oh, Billy!" The next minute, she was cracking jokes. I will never forget her saying, "Billy's probably going to go to hell, so maybe I should have his ass cremated." Penny and I were shocked, saying that was a terrible thing to say, but when she started laughing we couldn't help join in. Later she said, "Well, maybe I'll not have Billy cremated. There's no need for him to burn twice." Neither Penny nor I could believe this, yet it was the Aunt Minnie we knew and loved.

Then she started crying again and said, "Billy always said, if anything happened to him, there was some money in the box." She got up, went back into their bedroom, came out with a key, went in the closet, and pulled out a lock box. It was obvious to both of us that she didn't know what was in the box. Still crying, she unlocked the box and turned it upside down on the cocktail table. Out falls three Tender Vittles cat food boxes stuffed with money. With rubber bands around each of ten fat bundles of tens and twenties, each bundle was a thousand dollars. Penny and I were stunned. So was Aunt Minnie. The crying stopped, and her eyes began to sparkle. "Billy always said he'd take care of me. He was going to buy me a house."

I told Aunt Minnie we had to go put the money in the bank. It wasn't safe for her to have it in her house since she didn't know who knew about it. Every day, Uncle Billy had gone to work and then to the bar up the street for his drinks before coming home. This was his routine. No telling who he might have told about the money. But Aunt Minnie refused to do anything about it until Albert came home on Monday, so that was that. I couldn't convince her otherwise.

I went back over that evening to be with her when the undertaker came. Here I was sitting on the couch, while Aunt Minnie sat in the Queen Anne chair not knowing how to play the grieving widow. I felt more sorry for the undertaker than I did for her. I know that sounds terrible, but you had to be there. One minute she was going through the crying and moaning, "Oh, Billy! Poor Billy! What am I going to do without Billy?" Then she would wipe her eyes with her linen and lace handkerchief, and the next thing out of her mouth would be something funny.

The poor undertaker would just drop his head. I could see how hard it was for him to keep his composure. Her response to everything he asked was: "I want Billy to have the best." Sammy the cat kept coming through the room, making this crying sound like a baby. It was as if he knew something was wrong. I'm sure he wasn't used to company, because no one visited them. His crying sound was sad.

Finally, the arrangements were complete. The undertaker assured Aunt Minnie that everything would be done to her satisfaction. When his kind manner brought on more tears, he took his folder and got out of there as fast as he could. I stayed awhile. Aunt Minnie wanted me to help pick out a suit for Uncle Billy. She must have brought out every color: black, brown, navy, gray, and they were all in plastic bags, several still having the tags on them. That was a surprise, because neither Uncle Billy nor Aunt Minnie ever went anywhere dressy, but you'd never have known it from either one's wardrobe.

Monday couldn't come fast enough. Gene went to the airport to pick up Momma and Daddy. They were totally wiped out because they had flown straight through from Tokyo. Of course, this was not the kind of welcome home they expected.

The funeral was scheduled for Thursday, so that gave them a little time to get over their jet lag. Also, Momma had gotten very sick in Singapore after eating some raw fish. They had to have a doctor for her. So she still wasn't tip-top. Both Momma and Daddy were pretty calm and handled Uncle Billy's death with concern. However, when I told Daddy about the Tender Vittles boxes, with ten thousand dollars, he had a fit about it not being in the bank. Daddy, an impatient, organized, thrifty, logical-thinking, precise, no-nonsense, spiritual person, could not tolerate his sister's carefree, happy-go-lucky ways. "Minnie doesn't have a lick of sense," he'd say. He wasn't home one full day before he took her to the bank and deposited the money, although not before she kept some to spend.

At the bank, they found out that Uncle Billy had a safe-deposit box, but it was only in his name. That meant that it had to go through Probate Court before it could be opened. He didn't have a will. This would greatly complicate things.

Aunt Minnie wanted to buy a few things, so on Tuesday I took her to town shopping. All the classy shops are on Fourth Street, so that's where we went. She bought Uncle Billy two shirts, each costing about forty dollars, because she couldn't make up her mind which one she liked best. That was a lot of money in 1976. I couldn't convince her that he could only wear one.

Then we went to Jenny's, a classy women's apparel store. Of course, they knew her, because that's where she shopped. She told everyone about Uncle Billy's heart attack and death, which would bring on the tears and the lace handkerchief. What a day.

The funeral was held at the funeral home because Uncle Billy didn't belong to a church. There was Daddy and Momma, Aunt Minnie's five nieces, nephew, and cousins from my father's side. There were also friends from the hospital where she worked and family from my mother's side. So there was a full funeral chapel.

Well, lo and behold, as we were filing out of the funeral home behind the coffin, there on the street stood Uncle Billy's daughter and her mother, Uncle Billy's first wife. I had completely forgotten he'd been married before. After all, Uncle Billy and Aunt Minnie had been married for thirty-one years. The daughter and mother just stood on the sidewalk and watched as his body was placed in the hearse and the family got in the cars. I found out later that she didn't even come in to see the remains. I wondered why they had come at all.

I guess there's a reason for everything, although we might not know what it is at the time. The week after the funeral, the daughter called Aunt Minnie and told her that if she was going to inherit something, not to call her house, but to contact her at her job. She then gave her the phone number for her job. We all thought this took plenty of nerve.

Several weeks after the funeral, Aunt Minnie hired the husband of one of her friends as her attorney since Uncle Billy didn't have a will. Daddy worked with the lawyer to help get her affairs in order. He did all the legwork, going to the courthouse, getting death certificates, tracking down insurance policies, and so on, which saved greatly on legal fees. This was fine with Aunt Minnie. She knew she could trust Albert.

Now, remember, Daddy had already helped her deposit what we called the Tender Vittles money. Finally came the day when they met the representative from Probate Court and her lawyer at the bank to open the safe-deposit box. Well, dear Uncle Billy had obviously taken from his postal check enough money for household expenses and pocket money and had saved the rest. The box was crammed full of cash and U.S. Savings Bonds, somewhere in excess of eighty thousand dollars. Needless to say, everyone was shocked.

When things were finally settled, Uncle Billy's daughter was awarded a hefty chunk of that money, and my Aunt followed her directions and called her at work. She went by the attorney's office and picked up the check.

I later found out that my Uncle sent money to her mother on a regular basis while she was growing up. Aunt Minnie also said, "She never so much as sent him a card." He had grandchildren he had never seen. It's strange to me that Aunt Minnie wasn't the least bit upset over giving his daughter the money. Her feeling was: "Billy did right by you and you did wrong by him. You always get more doing the right thing in the end."

All of a sudden I realized I was pulling into my driveway. I was scared to death. I had been driving for thirty-five minutes, and it was all a complete blur. I remembered leaving the hospital and getting on the expressway, and then there I was in my driveway. My gospel tape was still playing, but it was as if I only heard one song: "His Eye is on the Sparrow." I pulled in the garage and just sat there, motor still running, tape still playing. I tried to get myself together before going into the house. It was January 1994.

five

January here is usually bitterly cold, with snow and tough weather. This year was no exception. We had to clean out Aunt Minnie's apartment, as well as get her settled in the nursing home.

During the weeks Aunt Minnie was in the hospital, Momma had been over to her apartment a couple of times to get things for her. She said to me several times, "Minnie's apartment is a mess. I can tell she hasn't felt well in a long time." This seemed strange, but no one visited Aunt Minnie on a regular basis. I talked to her on the phone, as did Momma, but I probably went by her apartment only a couple of times a year and Momma stopped by maybe every other month. Other family seldom visited or called. It seemed Aunt Minnie only kept in touch with us. Daddy seldom talked to her on the phone. He relied on Momma to keep in touch. However, if she needed money, then Daddy was on her case.

About six months after Uncle Billy died, Aunt Minnie moved into a two-bedroom apartment on the top floor of a large building. She had a balcony overlooking a golf course. She bought a new dining room set, bedroom furniture, a piano, and decorated the apartment very tastefully. So we're talking about her living in this apartment for over eighteen years.

I guess her money lasted about five years. I have to

admit, when she had it, Aunt Minnie was the most generous person you could find. She bought expensive gifts at Christmas. She helped my cousin by letting her stay with her for a while, even bought her a car. She gave all kinds of things to people in her building. She became active at our church, too. During this time, Aunt Minnie would call a taxi and go shopping and anywhere else she wanted to go, except she still didn't visit family.

So Thanksgiving 1986 was special because Aunt Minnie came to our new house for the family dinner. She had an absolute ball and kept everyone in stitches. "Lord child," she said, "You'll be paying for this house until Jesus comes. Then He's going to come back and ask for the rest of the money." This quip was classic Aunt Minnie.

Daddy was constantly talking—lecturing—to her about being so carefree with her money. The first couple of years, she had Daddy file her taxes. As she spent more and more, she then paid to have them done. If my memory serves me correctly, she went to Atlanta on a bus trip with some friends and on a couple of day trips to Kentucky. This was probably the only traveling she had done, maybe in her life. She bought a complete set of Invicta leather luggage and had custom clear plastic, zippered covers made for each piece. Then each piece was placed back in its original box. I think she used this luggage on the trip to Atlanta, but I'm not sure. What is interesting is that she never discussed this trip to Atlanta with me, though we talked regularly.

Aunt Minnie joined the Usher Board and Circle Nine at our church. She looked majestic in her fine black suits and white silk blouses as she walked up and down the aisle. She enjoyed the participation and especially loved entertaining her

Circle. After a while, she even became the chairperson. It was as though this woman who had lived in a cocoon had broken out, was blossoming and fully enjoying life. These had to have been good years for her.

During this time, she was most like the Aunt Minnie that Marsha and I remembered from our childhood. It seems as though we only saw her and Uncle Billy on Christmas Day. They would come over to our grandmother's, dressed to the nines, and bring us each a beautiful, expensive, knit dress. Although she didn't visit often, she lit up the house with laughter, and grandmother was always thrilled to see them.

The summer of my freshman year in college, Aunt Minnie got me a job at Good Samaritan Hospital with her. She marched into the Personnel Office and told the Sister, "This is my niece! I want you to give her a job for the summer as a Nurse Aide, and I want her to work the same shift I work." Since "Minerva" knew everyone and was well thought of by her coworkers, I started two days later, working 2:30 to 11:00 in the newborn nursery. This summer work experience really bonded us. I picked her up at home and dropped her off each night, and was known in this big hospital as Minerva's niece. I went back there to work the summer of my sophomore year.

I remember vividly how, each evening, after visiting hours, one of the nuns would say a prayer over the intercom. It was at nine o'clock every night. Then after the prayer, she would say:

"Four Angels at Your Bed,
Two to Foot and Two to Head,
Four to Carry You
When You Are Dead.
Good Night!"

Now, in the two years I worked there, I heard this every evening. I never asked anyone about it, but I always thought it was weird. I thought it was a terrible thing to say to patients in a hospital before they go to sleep. It seemed strange. The prayer was fine, but the Angel part was on the morbid side.

Toward the end of the first week of January, 1994, Aunt Minnie was transferred to the nursing home. Because of my work schedule and the number of days I had been off, I was unable to go with her. According to the nursing staff, she arrived in the early afternoon and was crying. When I arrived a couple of hours later, Momma had already been there. Aunt Minnie was in a room with two other women, right across from the nurses' station and near the recreation area for that wing. She wasn't crying, but was so sad, only answering what I asked her. She showed no interest in the other two women. She had withdrawn into herself. I tried to get her to go out and watch the large screen TV. After all, she was used to watching from morning until late at night, through Johnny Carson, doing her fabulous needlework. Her handwork was flawless, embroidering tablecloths, napkins, towels, all of course made of linen. But, now, she wouldn't budge from her bed. She was depressed, and I felt guilty.

On my way home, I cried and cried. I felt I had dumped her in this strange place and broken her spirit. When I got home, Gene tried to console me. He said he would go by and see her. She was very fond of him and asked about him every time we talked. I called Momma. She was reassuring, too, and encouraged me to calm down. As always, she found something positive to focus on, saying, "As far as nursing homes go, this is a nice place. We can all be assured that she's safe.

She'll get her meals and get good care." This was certainly true, but I couldn't wait to get back over there the next day.

It was the second week of January, and we hadn't done a thing to her apartment. So we began what was to be, in my words, twenty days of pure Hell. Every evening after work, Momma, Gene, and I would meet at the apartment at five o'clock and work until about eight-thirty. The first day there, I was ready to lock the door and leave the country. Momma's description—"It's a mess"—was so understated it wasn't funny.

When we went in the door, the living room looked fair, but certainly not up to Aunt Minnie's standards. The dining room table was full of newspapers, bills, magazines, books, you name it, all stacked on the linen tablecloth. The living room had the usual sheets on the furniture. Right by the hallway to the bedrooms, sat her Queen Anne chair. This is where she would sit when you visited, so you couldn't get back to the bedrooms. Books were stacked on the floor. There were cigarette butts in the ashtrays. Everything dusty. However, the shocker was that all of Aunt Minnie's plants were either dead or dying. She had at least twenty-five plants that sat in front of her balcony doors. They were usually healthy and full of blooms. We stood there just looking at what used to be her joy.

This in itself was a major sign that Aunt Minnie hadn't been well in some time. She took great pride in growing and caring for her plants. One of her hobbies was growing orchids. They were not only beautiful, but her source of pride. In fact, if there is such a thing as a green thumb, she had two. In the spring and summer, she grew all kinds of plants in pots on her balcony. I remember so clearly fussing with her about growing a marijuana plant there. Although she was on the ninth floor, the district police station was right next door. "They are going to come up there and get you," I said, "And you'll be calling me from jail." She would just laugh and say, "If they come over, I'll give them some." Of course neither Daddy nor Momma would have found this funny.

By this time, Gene and Momma had been through the entire apartment. Momma was saying, "I told you it was a mess." Gene was saying. "Oh my goodness! Oh, my!" As I got to the door of the first bedroom, I went into total shock. I couldn't get into the room. There were clothes, bed linens, blankets, boxes, books, an ironing board, a grocery cart from the lobby, the room all crammed full of stuff. The orchids in the window were dried up. This room was her den, with a piano, a gorgeous antique love seat, a large stuffed chair, and television. But it looked like a tornado had gone through. She had evidently just piled things here after doing laundry, rather than putting them away.

It didn't get better. In her bedroom, the bed was stripped down to the mattress, bedding on the floor, plastic bags filled with soiled clothes, her small black and white TV sitting in the middle of the bed, the dresser piled with ... all I can think of is "stuff," lamp on the floor, chair piled high with clothes. There was an indescribable odor in the room and throughout the whole apartment. It wasn't urine but a foul smell. We opened the windows and sliding doors even though it was ten degrees outside.

We had come prepared to clean, with our Clorox, rubber gloves, trash bags, and all, but there was no way either Gene or myself were psychologically prepared for this disaster, or Momma either, all of us finally realizing that Aunt Minnie hadn't been functioning in a long time. There's no way I would have believed it if someone had told me about this mess. Not my Aunt Minnie, earthy and carefree, but prim and proper, too. This woman who had to have the finest linens with lace, yet here was her bed stripped to the mattress, her sheets, blankets, and spread crumpled on the floor. I will never forget how sad this was, for each of us. It was something you think happens only to others, to those who have no one to care for them. But this had happened to my Aunt Minnie. How could she have covered, so skillfully, that she needed help?

Suddenly I realized I hadn't looked in the kitchen. There were dishes and burned pots and pans in the sink. The food in the refrigerator had to be thrown out. Her freezer had chops, steaks, chicken, juice, and frozen vegetables. She had all the necessary staples. Behind the door, she had stacked cases of Pepsi, and in the midst of the papers, magazines, mail, and mess on the dining room table, was a carton of Virginia Slims. Aunt Minnie, like myself, was a smoker, and as we would later find, had burned holes in much of her clothing—another sign that she was in need of help.

Now, the funny thing about all of this is that prior to her going into the hospital in December, I had talked to her on the phone almost every day and Momma talked to her at least once or twice a week. It never failed that I would ask her what she had for dinner. She would tell me something like, Oh, I had the best baked pork chops, a baked sweet potato with butter and sugar, and some baby lima beans with butter. That's exactly how she would describe a meal. And I'd say, "Wow, I wish I had been there," and we would laugh. The reality is that she hadn't been cooking for some time, and most of her pots and pans were badly burnt. It still blows my mind how she fooled us all into thinking she was fine.

It was already getting dark, and we hadn't accomplished anything. Momma made us close the windows and doors because it was freezing. Poor Gene, who cannot stand odors, had to hang in there. So, Momma started in Aunt Minnie's room, Gene started in the den, trying to clear a path so he could get in, and I started in the living room. We each had rubber gloves and started picking up debris and filling trash bags. Fortunately, the trash chute was just down the hall.

I frantically picked up newspapers and magazines in the living room, stuffing them into trash bags. Do you remember Aunt Minnie's personal relationship with Publisher's Clearing House? Well, you name any magazine on their stickers and she had it. She had the Time/Life series on Gardening, cookbooks, *Science, The Bible, Needlework,* and on and on. She got two copies of *Jet*, two copies of *Ebony*, two copies of *House Beautiful*, two copies of *Architectural Digest*, and the subscriptions didn't expire for another two, three, even four years. On the side of the couch, on top of the sheet covering it, was a stack of magazines as high as the couch arm.

I don't know how I must have looked, but Momma came in and said, "Carol, slow down. We can't do all this in one night, and you need to be careful about what you're throwing away." I sat on the couch next to the magazines and smoked a cigarette. I opened the doors to get some more of that cold air. I wondered how in the hell we were ever going to make a dent in all this stuff. After a few minutes, I began taking the trash bags to the garbage chute. The opening to the chute was so small that I had to unload the bags by hand. Momma cleaned off the dining room table, taking home all the mail and Aunt Minnie's purse and keys. I took home several bags of soiled clothing and started in on what was to become a nightly ritual with my Tide and Clorox. I didn't want Momma doing any of the heavy laundry, though she would later wash all the silk scarves, lace handkerchiefs, tablecloths, and kitchen towels.

As we left the building that cold January night, Momma got into her trademark Buick Century and drove off. Gene and I pulled out of the parking lot and headed home, both of us too tired to talk. All I could think was, How was I going to find the energy to get through this?

Once home, I put in a couple loads of laundry, got ready for the next day's work, nursing home visit, and apartment cleaning. I was totally, and I do mean totally, wiped out. Looking back, I can see that this ordeal was another example of the Lord "giving you the strength when you need it, not a minute before, but when you need it." I truly give Praise and Glory to the Lord for helping each of us, Gene, Momma, myself, and of course Aunt Minnie, get through the month.

six

It was the last day of January before we cleared everything out of the apartment and basement storage area. The woman across the hall took Aunt Minnie's plants. She replanted them, and they were again thriving. That made me feel good. She and her daughter, in sub-zero weather, helped us take cart loads of things to the outside dumpster. The apartment manager had told us not to worry if we weren't finished by the thirty-first, but when we left there that night, I didn't want to face another day in that place. Thank goodness we were finished.

The neighbors had all been very kind during the month, asking about Aunt Minnie and expressing how much they would miss her. After all, she had been there seventeen years and had kept up with everyone moving in and out.

All during this month, when I would visit Aunt Minnie, I would tell her something about cleaning out the apartment. I told her we had moved some things to Momma's and her furniture to my house, but she seldom asked any questions and would pretend she didn't care about her things. She would always say she wanted Kim to have her furniture, maybe because she knew Kim was getting married, but she mostly blocked out what was going on and didn't want to discuss the apartment.

The nursing staff was concerned because they could seldom get her to come out of her room. She refused to eat in the main dining room and insisted that her meals be brought to her room. She wouldn't go out and watch television with the other residents, nor go to the lounge for the activities. Even I couldn't get her to do these things. She complained about one of the women in the room constantly trying to talk to her, and since Aunt Minnie could not understand or accept this, she fussed at her constantly. So during this first month in the nursing home, I was the referee between these two. Then I had to listen to the nurses and assistants tell me about Aunt Minnie being stubborn and how she wasn't trying to do things for herself.

January 31 was Momma and Daddy's anniversary, and despite all that was going on, Marsha, Gene, and I decided to get them a 13-inch color television for their kitchen. Their old one was all fussy, and the picture sucked. Now, it would never dawn on them to go buy a new one, so this was the perfect gift. Gene and I got the television, went over and put it up on the shelf, and the picture was excellent. Momma was thrilled, and Daddy, of course, went into one of his lectures about not buying everything you see. He would never admit it, but with his limited eyesight, he had to see the picture better.

February 11 was Aunt Minnie's birthday. I told Gene maybe we should buy a television for her, too, for her room. I talked to the nurse and told her what I was going to do. She wasn't thrilled. She seemed to feel that they'd never get her out of her room if she had a TV. I gave it much thought. I knew how much she loved watching television—that was how she kept up with the news. So we bought another 13-inch color television. The salesperson laughed and asked if I was sure I didn't want any more. It was only the week before that we had bought Momma and Daddy's.

On Aunt Minnie's birthday, Gene and I walked into her room at the nursing home with the television. I tell you, I will never forget how happy it made her. She was sitting on the side of the bed, a Happy Birthday balloon tied to the headboard, a large cupcake with a candle sitting on her table, and was looking down at the floor. When we both said "Happy Birthday!" she broke out in a gigantic Aunt Minnie smile.

"Is that for me?" she asked.

"It is if you're the birthday girl."

Aunt Minnie scrambled to her feet, started clearing her table, and was about to throw her cupcake in the garbage. I grabbed it, while Gene put the television on the table. He plugged it in, and Thank the Good Lord, the reception was fine. She could get all three local channels, plus the educational channel, with a good crisp picture.

Leave it to Aunt Minnie to say, "That's great, but can I get cable?"

"No," I said. "If you want to watch cable, you have to go out to the lounge."

"That's OK. This will be just fine," she said with a big grin.

Aunt Minnie is a trip!

When we left that evening, she walked us all the way to the front door. This was the first time she had done that, and I can't think of another visit from then on when she didn't walk me to the door. In warm weather, she would walk me all the way to the car.

It seems that after Aunt Minnie got that television, things began to turn around for her. She continued to fuss with the other woman in her room, so they moved her up the hall into another three-bed room. Then it was Minnie, Lorena, and Miss Beal. I must admit they were much more compatible, because they were able to communicate on the same level. Although diagnosed as having mild dementia, Aunt Minnie, to me, had bounced back to her old self. She turned her television on early in the morning, and it stayed on until late at night. It didn't seem to bother the others. She was back in the saddle with all the latest news. Though the nurse was concerned that the TV would make her worse about participating and leaving her room, it had worked just the opposite.

At last, Aunt Minnie began to adjust. She ventured out into the hallways, lounge, and dining room. She quickly knew everyone's name, who had visitors, and how long patients had been there. She had gotten in with the staff, from nurses to kitchen workers, and she told them what she liked and what she didn't. They would let her make calls from the nursing station, and she would call me or Momma sometimes, usually wanting me to bring her something.

Momma would go by and visit her every Wednesday after her bowling. I went three or four times a week after work. I did her laundry, so I had this large yellow plastic tote bag. All the staff and patients began to know me as "Minerva's niece." (They called her Minerva.) I always had the yellow bag in the back of my car with either clean or dirty clothes. This way I went in with the clean and came out with the dirty clothes each visit.

Aunt Minnie was now busy shuffling around every day, playing Bingo, singing with the choir, serving on the resident council, visiting with the staff, and welcoming new patients. She was dressed every day. No longer was she sitting in her gown and robe on the side of her bed. Everyone, and I do mean everyone in the place, including visitors, knew her. My old Aunt Minnie was back!

Off the foyer was a big round planter that was maintained by a landscape company. It was changed every couple of months with various seasonal plants. Whenever the people came to tend this planter, Aunt Minnie was out there to direct them and would tell them if she thought something wasn't healthy. After all, this is the lady with two green thumbs.

By early spring, Aunt Minnie was back in full bloom, so I began to feel better, and less guilty, about assuming the responsibility for her being in a nursing home. When I completed the paperwork with the social worker in January, one of the questions was whether she had a Living Will. Of course, the answer was No. Then the social worker asked if I wanted a "Do Not Resuscitate order" on her chart. I remember being startled, tense, then forcefully saying, "Of course not!" I wasn't prepared for the question, and I'm sure my reaction showed this. The lady assured me that this was required on all charts, along with the name of the funeral home the family chose to use. Boy, I walked out of there that day, thinking, "Shit, I can't deal with all this. After all, Aunt Minnie wasn't sick, you know! She just needed someone to look after her since she wouldn't come live with us." Thank goodness that was all behind us.

Two of the women from our church, Mrs. Harris and Mrs. Taylor, would come to visit Aunt Minnie. She entertained them and kept them laughing, as well as getting all the details about what was going on in the church. Believe it or not, Momma got Daddy to go visit a couple of times. He hated the idea of a nursing home, but he had to admit she was doing just fine. Daddy is always so serious about everything. Since losing most of his sight, he just can't seem to tolerate humor. That means Aunt Minnie was busy telling everyone that this was her brother Albert and cracking jokes. This made Daddy so irritated that, of course, he had to say again, "Minnie doesn't have a lick of sense."

Now, not only did Daddy dislike the idea of a nursing home, he made it abundantly clear, anytime the subject came up, that he believed in life support. His famous words are: "I do not want anyone pulling the plug on me. When my time comes, it will be The Lord's decision." Momma and all of us would say, "Don't worry, Daddy. We know you don't want anyone to pull the plug." These discussions would usually end with something light being said that would make even Dad smile or laugh. But he was truly serious about wanting to be kept on life support, no matter what.

Aunt Minnie, being such a television junkie, was into every detail of the murders of Nicole Brown Simpson and Ronald Goldman in Los Angeles. She couldn't talk about anything else. She was positive O.J. was not guilty. She had all kinds of theories and had everyone in the nursing home, mostly staff, listening to her opinions. If they disagreed, she would launch a mighty case equal to that of Johnny Cochran.

Well, of all times for her to be in the lounge watching television, she just happened to be there when the Bronco chase occurred. When I arrived the next day, there were several people sitting out front under the canopy. I would always greet them as I walked into the building. But this time one of the men said, "Wait until you hear about Minerva." I smiled and went through the door. Within seconds, here's one of the nurses saying, "Minerva led a cheering session last night and had all the patients in the lounge saying, Go, Juice, Go!" It was obvious she wasn't pleased. Needless to say, there wasn't anything I could do. I just told her that she knew what a character Aunt Minnie was and that I hoped it didn't cause too much of a problem. I said all of this with a straight face and serious tone, and she didn't make any further comment.

I went into the dining room, and there sat Aunt Minnie at her table in the center of the room. She looked up, and with that big devilish grin, said quite loudly, "Carol Ann, guess what? Last night I got everyone to cheer for O.J." By this time, her table mates and some of the other patients were telling me, "That's right, she did." Aunt Minnie was just thrilled, giggling and laughing, getting all this attention. I wouldn't discuss it and didn't ask any questions. I just told her to finish her meal so we could go to her room. Her immediate response was, "I'm finished," and she proceeded to get up, still talking about O.J.

On my way home, I became hysterical. People in passing cars probably thought I was nuts! I had this picture in my head of these elderly patients, some in wheelchairs, some disoriented, with Aunt Minnie up and leading this chant, "Go, Juice, Go!" Boy, that's my Aunt Minnie! She could find the bright side of anything, and I do mean anything. There was no way she'd ever be taken for a shrinking violet. That's what made her so lovable.

seven

Through all of this, Momma maintained her busy schedule. She was truly a person who had to be involved and doing things that were constructive and would help others. Unlike Daddy, she didn't allow Aunt Minnie to get on her nerves. She was always very kind but direct with her, and Aunt Minnie respected whatever Harriet said. She would challenge Daddy because she knew she could get him agitated. This is the way it had always been.

Momma had her weekly schedule, and like the letter carrier, rain, sleet, or snow couldn't stop her. Of course, this would keep Daddy fussing because he didn't want to leave the house and wanted Momma to stay in also. Momma would fix his breakfast and lunch and then take off for either bowling, golf, sewing group, church, the shopping center, or wherever. If anyone was sick, then she added a visit to her schedule, either at the hospital or their home. You could depend on Harriet, and everyone knew this.

Then there were monthly meetings for the Optimist Welfare Club, where she was treasurer, and the Wilberforce Alumni Club, again treasurer. And, of course, Momma was at church every Sunday. Most of the time she would get Daddy to go, too. Oh, I almost forgot Momma's church group, Circle Six. She was Chair and had been for years. They also met monthly, and as with the other groups, if they didn't have a place, then Momma volunteered.

This makes me remember when Kim was growing up. Her favorite thing was to help Momma entertain her clubs. Kim has been an outgoing, friendly charmer all her life and would see to it that the women were thoroughly entertained. When she went away to college, all the women wanted to know about Kim. That's when I had to take over and help with serving and cleaning up. It wasn't the same for them, because they preferred Harriet's granddaughter.

Momma was the sixth child of twelve. I don't know what significance that holds, but she was the one that all her siblings depended on for everything. When my grandmother got sick after my grandfather passed, it was Momma who took care of her. She came to live with us and died at our house on Harvey Avenue.

When Momma and Dad were first married, several of her sisters and their families lived with us until they were able to get out on their own. We lived in a big old white house on Blair Avenue. Daddy had bought it before he and Momma were married. His mother, brother, sister, and nephew lived on the first floor, and we lived on the second floor and had a playroom on the third. This was a wonderful family home, and Marsha and I were truly spoiled by our paternal grandmother. There were several cousins, from both sides of the family, born while we lived on Blair Avenue. Their parents were staying with us at the time: Jackie, Jeannette, Lorna, Lynne, Michelle, and I think Linda. Anyway, it was a wondrous home and Momma was the center, along with Grandmother Walker.

We lived there until I was in the fifth grade. Then we

moved to Harvey Avenue, but even on Harvey, anytime someone came from out of town, it was understood that they would stay with us. Our home was like the Holiday Inn, and we all enjoyed every minute of the visits.

Momma was the main person who cared for Aunt Mary, Daddy's sister, through her illness and death of cancer. Aunt Mary left two daughters, ages five years and six months. Momma helped grandmother and Mary's husband Lloyd take care of the girls. About three years later, Lloyd remarried. Shortly after that, grandmother was diagnosed with cancer, and Momma, Daddy, and Uncle Howard, Dad's brother, cared for her until she passed. This was truly devastating for me, because I was so close to my grandmother. She was always so proud of me. I was a sophomore in college when I made the train ride home for her funeral. In my mind, it's as if it was yesterday, but it was January 1962.

At the time Grandmother Walker died, I didn't get in until late the night before the funeral, so I had missed the wake. I vividly remember walking into the Seventh Day Adventist Church behind the family and sitting in the second row. The coffin was open throughout the service, and toward the end of the service, they had that "Parting View" for the family. When the second row stood to file out, I started pushing everyone so I could get out. As I approached the coffin, I saw this frail, white-looking woman, with snow-white hair. I cried, loudly, "That's not my grandmother!" Sobbing and repeating this, I suddenly had Momma on one side and Daddy on the other. They had gotten up from the front row. Momma whispered, "Carol, calm down. That's the way grandmother looked when she was young." Daddy had his arm around me. I don't know what I said, but I turned, walked out of the sanctuary, and stood in the lobby crying. The ushers tried to console me, but I was consumed in my own thoughts. Reverend Wright, the pastor from our church, Mt. Zion, came out and talked with me. All I could remember was that the last time I saw Grandmother in the hospital, the very last thing she said to me was, "I never taught you to make cornbread."

As the years passed, Momma was there for her brother Bud and sisters Clara, Edith, and Ruth, her sisters passing with cancer, Uncle Bud with diabetes. She was there for her aunts, Addie Belle and Ruth, as well as for her niece Barbara. Then there were relatives on my Daddy's side of the family: Uncle George, Uncle Howard, Aunt Evelyn, Aunt Mary, and of course Uncle Billy. So you see, everyone truly did depend on Harriet.

Now, not only was Momma a part of the sad times, but she was also always in the center of the weddings, christenings, family picnics, and anniversary celebrations, and the family took pride in sharing their good news with her. Remember, each Christmas your holiday calendar had to include Harriet and Walker's Open House the day after Christmas. This meant young and old.

eight

Momma had given me some slack since we had been trying to get Aunt Minnie straightened out, but her primary concern had shifted to Kim and Toren's wedding. They had decided on Labor Day weekend in order to make it convenient for Toren's family to come from St. Louis to Cincinnati. September 3 seemed a long time away. But it was now March, and even *I* knew I had to face up to this wedding.

By this time, Kim and Toren were looking at houses. I certainly encouraged this, because a mortgage wouldn't be much more than rent. Toren had a daughter, Tiffany, who was about ten years old and wanted to come live with them. Kim accepted Tiffany just as if she were her natural daughter, so this presented no problem. I was proud of the way she related to Tiffany and yet didn't try to replace her mother.

After long discussions around our kitchen table, Kim and Toren had decided to build a house in the suburbs, 30 miles out of St. Louis. It was a new subdivision, and they (Kim) had selected the largest model. The loft outside the second-floor bedrooms was what sold Kim on the house, because she envisioned putting her drafting board there, along with all her crafts and photography stuff.

Now, of course, comes the money part. They had enough income to qualify for the house. However, as you might guess, they didn't have enough up-front cash. That's where her Dad and I, as well as Momma and Daddy, came in. Our proposal was that we could help with the down payment, if they would have a small, family wedding. Both Kim and Toren agreed and, within a couple of weeks, had signed on the line. The house was to become a reality and would be finished by late fall.

Once this was all settled, then came the plans for this "small family wedding." Daddy didn't believe in wasting money on weddings, so he was pleased with their decision. In her own way, Momma was as thrifty as Daddy. She shopped daily for bargains. But Momma, although she didn't say so, looked forward to Kim having a big wedding. I could tell. After all, this was their first grandchild.

Well, Kim, Gene, and I had a huge argument over what constitutes a "family wedding." I maintained that if we invited one person who wasn't family, we were setting ourselves up for problems. Everyone we knew liked or loved Kim, because she had grown up embracing and caring about people. Whether they were young or old, Kim always had time to brighten their day. So we're talking about my kid knowing everyone, for she met no strangers.

I had taught Kim, as I had been taught, to call close friends of the family Aunt or Uncle, out of respect and to signify closeness. Well, this all backfired on me, because Kim started rattling off names, saying we must invite, Aunt ... and Uncle ... and Uncle ... and on and on. They both argued that we couldn't exclude Gene's friend Uncle Jack or Mom's friend Mrs. Wilson and so on. I yelled, "They are not family!" Then Gene chimed in, "Now, Carol, you know there's no way we can't invite them. That would be terrible." As had always been the case, Kim and Gene, two peas in a pod, had ganged up on me, and I was bound to lose this one hands down. I conceded, but not without stressing the fact that we had just shelled out for the house.

Kim, the charmer, said, "Mom, Toren and I really appreciate the help you and Dad have given us, and we'll keep things as simple as possible. Anyway, we'll have to depend on you to do most of the planning because we aren't here." She also knew that when I plan something, I put it together right, because I'm a detail person. The year before, I had coordinated her cousin Heather's wedding, and I must say, it was spectacular. In fact, one of the attendants asked Heather, "Was your Aunt in the military?" So Kim knew she didn't have to worry. Without admitting it to either her or her Dad, I knew my baby would have a wedding to remember.

Of course, this just added more to my plate. I had to work, look after Aunt Minnie, check on Momma and Daddy, and now plan a wedding with no time to spare. My first job was to secure a place for the luncheon reception that would follow the 12:30 wedding. Now, if there's one thing I know how to do, it's how to negotiate a deal. My theory is, as my Daddy taught me, "Nothing ventured, nothing gained." All they can say is no. I'm not the least bit embarrassed, and I get my way more often than I get turned down.

I chose a restaurant across the river from downtown, The Conservatory. I instantly knew it was the right place: beautifully decorated, with a huge atrium, tropical plants, three levels for dining, a bar, a dance floor—definitely the right place. Now to negotiate, and I do mean negotiate, everything I wanted. After all, the house money was supposed to be our gift to the kids, and the wedding was supposed to be small. Yeah, sure.

As a family everyone was helping to get this event together. Kim was responsible for getting her dress. I had to find the bridesmaid dresses. Momma and Daddy would give her the wedding cake, made by our Aunt Mary Lou. Marsha insisted that she pay for the flowers. And Aunt Minnie was busy reminding me that she wanted Kim to have her furniture.

Now the shock. Between the lists made by Kim and Toren, Toren's mother, Momma, and me, we were at 375 invitations, and most were for at least two. This threw me into total panic. We took an ax to the list, and I finally ordered 350 engraved invitations, knowing most of the out of town people wouldn't come.

I booked hotel rooms and arranged for my cousin Eric to cater the Friday night rehearsal party and the Saturday night party at the hotel for the out of town guests. I told the restaurant how I wanted tables decorated, bought dresses for the bridesmaids, and practiced "Here comes the Mother of the Bride" in my fabulous, new, hunter green suit.

When Kim and Toren came home to register, she turned up her nose at the bridesmaid suits. I flipped my lid—she doesn't like to dress up for anything! But this time I prevailed. They would wear these suits with black patent pumps and white hose. Kim finally conceded they weren't *that* bad but insisted that she at least wanted to pick her flowers. She had insisted all along that she wanted a round bouquet, not a trailing one. Fine. Trying to keep my blood pressure down, I sent her to the florist with her father. Gene later said she made it very clear what she wanted.

Things slowly fell into place. Kim had her dress, and my final job was to settle the details for the reception, everything from the menu to champagne, to where to put the table for gifts and a stand for the guest register.

Although Aunt Minnie had talked about this wedding for months and had shown everyone in the nursing home the invitation, she still refused to attend. One of her favorite attendants, Rosa, had even said she would take her and bring her back whenever she was ready. Even this didn't work. She just wanted to see the pictures. There was no reason health-wise that she couldn't go, so finally Momma told me to leave her alone.

The week of the wedding was terrific, even though I was a nervous wreck. Stacy, Kim's good friend from college, came from California and spent the first part of the week with us. The schedule for the week was: Thursday night, dinner for Toren's parents at our house; Friday night, rehearsal at the church and then dinner at our house for the wedding party and out of town guests; Saturday, the wedding, reception, and a nine o'clock party at the hotel; and Sunday, gift opening at our home.

My first surprise was on Thursday afternoon when Sondra pulled into my driveway. We had gone to Ohio University together and hadn't seen each other in at least twenty years. We jumped up and down squealing, laughing, and hugging each other. Stacy and Kim were in hysterics, because I had been teasing them all week about all their giggling. Now it was my turn to be teased.

Soon people started coming in: Momma and Daddy, Marsha, Kristy, Kevin, Michael, Norma and Luke, Toren's parents, and of course Toren and his friends Mark and Mark. I couldn't seem to get things organized because of talking to Sondra and everyone else. Gene had grilled steaks, and

Momma, in her usual manner, went about getting things in serving dishes and pulling the meal together. With Gene and Momma's help, we threw a hell of a dinner. The bride-to-be was no help, because she was busy being her social self, entertaining everyone, but that evening was so comfortable for all of us.

Every day, as the United Parcel truck pulled up, I became more and more amazed at people's generosity. These kids had received almost everything on their gift register. Gene and I often said that the response was a real tribute to the way people felt about Kim. I was deeply touched and Momma was thrilled. It was as if she was living out her dream through her granddaughter's wedding.

On Friday, the pace picked up. The fellows had to go pick up their tuxedos, the girls had to get their hair done, and people were arriving all day at the hotel. There were two conversion vans full from St. Louis, along with several cars. My good friend Harriett, another Ohio University classmate, came in from Cleveland, saying that another classmate, LaWanna, and her son Brandon were also coming. This was just wonderful, because I hadn't seen LaWanna in twenty years either. I heard from Harriett more often, but hadn't seen her in about ten years. So the wedding was turning into an Ohio University reunion.

In the afternoon, I went by the nursing home and made one last attempt at talking Aunt Minnie into coming to the wedding. I told her what she could wear and how we would arrange everything. Without hesitating, she told me flat out No! She was so excited and everyone in the building knew that Kim was getting married the next day, but she let that reclusive side of her personality dictate that she not attend. I really felt sad, because no one enjoys finery more than Aunt Minnie. She walked me to the car and kissed me and told me to bring her some cake and a program. As I drove off, I glanced back and she was waving and blowing kisses. I waved back and was on my way.

The rehearsal at church was at six. Everyone was there, and Dr. Clark, the pastor, was very efficient in getting the wedding party together and settled down. The wedding service was to follow the Methodist ritual, and he liked to have a sermon directed to the couple. I wasn't sure how this would go, but trusted that he knew more then we did. Dr. Clark was a newly assigned pastor to our church and had only been there a month. Although I had met with him several times, he didn't really know our family. But Kim and Toren really liked him, and that was all that counted.

After the rehearsal everyone took off, and by the time Gene and I got home, our street was lined with cars. At least fifty people must have been there for the rehearsal dinner. It gave the St. Louis people time to mingle and feel welcome and a part of the celebration. The two vans from St. Louis, some of Toren's relatives, were lost and didn't ever show up. No one missed them until late.

Also, no one realized that Toren and his friends Mark and Mark had disappeared for a while. Then I heard Kim scream, "Oh, no!" and everyone turned to see her looking at a stack of pictures. Toren was beaming. The pictures were of Toren standing on a pile of dirt with his arms outstretched and of him pointing to a large hole in the ground. It was the hole being dug for the foundation of their new house.

The week before Kim came home, she and Toren had been going to the lot every day only to be disappointed that the builders hadn't started. But before Toren and his friends left town on Thursday, they went by and were able to take these pictures. Then, while the party was going on, Toren and both Marks went up to the Photo Shop. It had just closed, but they banged on the door, told this long story, and persuaded the man to develop the pictures on the spot. It was truly a thrilling moment for Kim and Toren. We all shared in their joy and excitement. This was the icing on the cake for this pre-wedding party.

nine

September 3 was the day of the wedding of all weddings. It was Kim and Toren's day, and yet it belonged to all of us. Kim's friend Stacy had gone to the hotel on Friday, so Kristy, Kim's cousin and maid of honor, spent the night with us. Everyone was to be at the church at 11:00 a.m., so I had to deal with everyone else's nerves as well as my own. Both Kim and Kristy were to dress at home. This wasn't a problem since she didn't have a long dress and train. Kim was trying to pretend she was cool, but I knew better.

Since this was a blessed day, Gene suggested we leave early enough to stop by and see Aunt Minnie. I called the nursing home and told the nurse to have Aunt Minnie meet me at the front door. I said I had something for her and would be there in about five minutes. When Gene pulled into the circular drive, I could see Aunt Minnie standing inside the front door. She saw the car and came out the door with a pink gladiola in her hand. She had gotten it out of one of the floral sprays in the lounge, brought in from a patient's funeral. Her face lit up like a Christmas tree when we all got out of the car. She was so thrilled she started crying. She hugged and kissed each of us and gave her stamp of approval to Kim's dress.

When I turned around, the whole front window and door was lined with nursing home staff and patients. Many of them were also crying. Gene took pictures of Aunt Minnie and Kim, Aunt Minnie and the rest of us, and pictures of her by herself. By this time, we were all pretty shaken, Aunt Minnie in her usual words saying, "Go on, now! Go on!" After another round of kisses, we were off. As we pulled out of the driveway, Aunt Minnie was crying and waving her gladiola at us, and the staff was still lined up inside taking in this wonderful moment that had simply made her day. It was a moment she would always remember. I'm sure she never forgot. I'll never forget that moment either.

I truly believe that The Lord smiles on those who give of themselves to others. On that beautiful Saturday morning in September, we took time to share our joy and happiness with Aunt Minnie. As we rode to the church, each of us expressed how the visit had touched us. I will never forget how happy we made her. Despite her refusal to attend the wedding, we had made her a special part of the day.

By the time we arrived at the church, I'm sure my blood pressure was sky-high. I had made it clear that this was one wedding that would start on time. That meant the music would start at noon and the processional at 12:30 sharp! Since I knew my daughter well, I had assigned one of our cousins, Clara, to help her get ready. Kim, as well as her cousin Kristy, tend to have a flair for the dramatic. I knew there would be no way my nerves could deal with her, nor would she listen to me. So Kim, Kristy, and Heather were downstairs in the Rev. Ruffin Room, and the men, Toren, Mark, and his brother Tony, were upstairs in the church parlor. Momma and Daddy, Marsha, Mr. & Mrs. Jones, and myself were to be seated in the rear of the church, on the center aisle. My cousin Lorna was coordinating the hostesses. This way, everyone was where they were supposed to be when it came time for the processional.

The church was beautiful. Huge white floral arrangements, each with a large purple and white bow, sat on tall pedestals on either side of the altar. Mr. Robinson had done a fantastic job decorating the pews with white satin bows and greenery. I looked around, and yes, I was pleased. The organist, Mrs. Robinson, started playing at noon. I realized then that this was it. Just as I began to relax, watching people being seated, someone came up and whispered, "Kim needs to see you downstairs." My heart pounded as I got up and hurried downstairs.

When I walked into the Rev. Ruffin room, there stood Kim in the middle of the floor holding her bouquet up in the air, crying, "This is not what I ordered! I told them I wanted a *round* bouquet!" Gene looked at me and said, "That's right. She told them." Well, this blew me away. I could have whipped them both. The bouquet was not round, but it was beautiful. Kim was crying like it was a bouquet of crabgrass. As I stood there in a daze, Gene said, "I tried to pull some of the things out, to make it look more round." Finally I said, very sternly, "Kim the flowers are fine. Pull yourself together. I'm going back upstairs."

I closed the door, went back upstairs, and was seated again. Momma wanted to know what was wrong. I told her about the flowers, and she laughed, saying "Kim's just nervous, and you know how dramatic she can be." We both laughed and I felt better, but when I walked out of that room, I could have taken her outside and whipped her with a switch. Momma was right again, though. That's all it was—she was the nervous bride.

Shortly, Marsha came up to take her place with us. She said Kim was over the flowers but that there was another problem. Marsha's daughter Kristy couldn't find Toren's ring. Marsha had evidently done the same thing I did: told her to find it and then came upstairs. Momma smiled and said, "Now, you know, Kristy is probably nervous, too!"

The order of the processional was: Seating of Aunt of the Bride, Seating of Grandparents of the Bride, Seating of Groom's Parents, and Seating of Mother of the Bride. Kevin and Michael, my nephews, were "clean" in their tuxedos and did the honors in escorting the ladies down the aisle and pulling back the runner. I've never seem them so serious, or maybe they were scared, but I teased them both about how well they cleaned up.

Because of Daddy's eyesight, Momma and Daddy walked down the aisle together. Momma looked regal in her two-piece raw silk dress and coat which she had made when they were in the Orient in 1976. She had been surprised she could wear this outfit. She told us she must have lost quite a bit of weight, as she hadn't worn it in several years.

The church was almost full, and at 12:30 on the dot, the processional began. When I turned and watched Kim and her Dad coming down the aisle, I held my breath. She was truly radiant. My beautiful daughter, The Bride. Gene was walking straight as a poker and had a slight smile on his face. What a magical moment and memory. The service was beautiful. I managed to keep it together until the soloist sang The Lord's Prayer. Then I started to cry. I felt such joy, or happiness, or something. "It was well with my soul!"

At the end of the ceremony, Reverend Clark said to Toren, "You may now congratulate your bride." So what does my crazy daughter do? She sticks out her hand for a handshake, then they kissed. Well, the entire church was hysterical! See. I told you Kim had a flair for the dramatic. Then, as they were walking up the aisle, she saw four coworkers from St. Louis—she hadn't known they were coming. She threw her arms up, waving her bouquet at them, and again the audience laughed. Our daughter The Ham! Poor Toren was serious as a heart attack through the whole ceremony. He was probably wondering if he was marrying a comedian.

What a wonderful wedding! What a Glorious Day, the uniting of two families with Blessings from Above.

ten

Following the ceremony, we had the customary picture-taking at the church, our friend Vertis the photographer. I was surprised that Momma didn't shy away from the picture-taking. Before Kim was born, Momma had a facial stroke called Bell's Palsy, which left her face a little twisted on one side. I guess we were all used to this, because it wasn't that noticeable to us, but Momma was self-conscious about it, and pictures often seemed to highlight the twisted side of her face. When Momma was caught without posing, the pictures were just fine. But the bottom line was that she hated to have her picture taken.

On this occasion, Momma was so jubilant she posed for all the pictures, laughing, saying, "Make sure you get my good side." We all had a good time, cracking jokes, teasing Kim and Toren about the ceremony, and just enjoying the moment. Believe it or not, that included Daddy.

The Conservatory was beautiful. As we entered the lobby, there was the magnificent cake made by Aunt Mary Lou. She had outdone herself with fancy details, scrolls, leaves, buds, all in white. You had to look closely to see all the detail. With the lush tropical plants in the atrium as backdrop, it was picture perfect.

People were already being seated for the luncheon, and

the piano player was doing his background-music thing on the grand piano. After eating, Kim and Toren cut the cake and had pictures taken, then went from table to table, personally greeting the guests. This went over big, especially since Kim is such a ham.

As far as socializing, whenever our family gets together, you don't have to worry about that. Momma worked that room just like Kim and Toren. Daddy stayed in his seat, but enjoyed everyone coming up to speak to "Walker." Gene, the other social person in our family, was really into being father of the bride and seeing that everyone had a good time. And I guess I was just tense and nervous. I made the rounds, but pretty much stayed with Daddy. I know one thing: I was still praying that everything went well.

Kim was disappointed that people began to leave after they finished their meal, but a lot of the guests were older and a wedding makes for a long day. As for me, I was exhausted and was glad I'd have time to go home and rest before that evening. In the end, Gene and I were both pleased with how it all went. Excellent service, excellent food, fabulous decorations—it was quite an affair.

Around three-thirty, Gene and I finally got in the car to head home. The out of town guests were going back to the hotel, and I planned on going home, getting out of those clothes, and taking a nap before the evening party. As we were getting on the expressway, Gene said, "Oh, I invited some of the club members to stop by the house for a drink." When I sighed Oh no!, Gene said, "You don't ever want to invite anyone over, and that's ridiculous!" I was too exhausted to say anything, so I just kept my mouth shut.

When we arrived home, a couple of cars were already

there waiting for us. I put on my Carol Smile and said, "Come on in!" Now Gene loves, and I do mean loves, to entertain. The more the merrier, as far as he's concerned. All I could think was how tired I was and that I just wanted to rest. But friends are friends, so I went into our room, changed into a comfortable pants outfit, got out of those new heels, and put on my bedroom shoes.

When I got downstairs, everyone was having a ball, and I must admit I enjoyed recapping the day's events with them. Everyone was cracking up about Kim shaking Toren's hand and the look on his face in that split second. It also made us feel good to hear our special friends say how nice the wedding was and how pretty Kim looked. I wasn't the only one who had gotten comfortable, either. All the women had taken off their shoes, and the men loosened their ties.

It was close to nine o'clock when everyone finally left for the hotel party. By this time, Kristy was back at the house and had gone to bed, saying she was too tired to go to the party. Me, too. There would probably just be young folks there anyway. When Gene and I got in the car, I told him I really didn't want to go. I was so tired. But he said, "You have to. They have a surprise for you."

As we pulled into the hotel parking lot, I felt like I didn't have the energy to walk from the car to the lobby. Once inside, I was shocked when I walked into the room. It was almost wall to wall people, the food already gone, the bathtub full of ice and beer. Young and old alike were having themselves a sure-enough party.

I stood there a few minutes trying to take all this in, when Sondra, my friend from O.U., said, "We have a surprise for you! Look who's here!" I looked right into this couple's face and, with no tact at all, said, "I don't know who that is." Sondra quickly said, "Bonnie and Vernon," at which point I squealed and hugged them both. Sondra and Harriett had looked them up in the phone book and invited them over. I didn't even know they lived here, but Sondra did. Now it was an official Ohio University Reunion, with Harriett, Sondra, LaWanna, and now Bonnie and Vernon. This meant so much to me, even though I didn't have much time to spend with them. Kim's wedding had brought us together after more than twenty years.

The guests from St. Louis had a good time, and of course, so did Kim's high school friends, her St. Louis coworkers, her good friend Ann and spouse from Kansas City and Ann's parents from St. Louis. Then there was Arlene and Michael who came from St. Louis, Toren's family, and his brother Tony from Orlando. All these people cared about Kim and Toren and were there to share in their joy. The excitement and happiness in that suite that night was something to see, and The Lord had Blessed This Day! The excitement must have revived me, because I hung around until past midnight.

The following day was my birthday. Since some of the St. Louis group had missed the rehearsal dinner, we invited them over, along with Toren's parents, to open gifts. This made for a nice family gathering before they got on the road. My friend Harriett wasn't leaving until Monday, so she spent the day with us, too. Kim and Toren were in and out, doing laundry, getting ready to leave for Jamaica the next day. They bought me a beautiful gold and onyx necklace for my birthday. Kim had given me the matching pin for Christmas.

It would be Monday before I could go see Aunt Minnie and fill her in on all the details.

eleven

The week following the wedding was almost as hectic as the week before. Monday I took time to go by and visit with Aunt Minnie. She was delighted and would call in everyone who passed by the room. Now, almost everyone knew about the wedding, especially since we had gone by before the ceremony. Some told me how beautiful Kim looked and how Aunt Minnie cried and cried, tears of joy, because her special niece had come by to share the occasion.

People were still dropping by the house with gifts, too. On Wednesday, I went back to work and enjoyed telling everyone about the wedding. Momma still hadn't come down to earth. She was so pleased to have heard from so many of her friends, all of them saying how much they enjoyed the occasion. Momma and Daddy went over and over every detail. Daddy was constantly saying, "I couldn't see," but believe me, he didn't miss much. When both asked if we had heard from Kim and Toren, I told them, "They can't be calling from Jamaica. They don't have any money." Then we would all laugh!

That weekend the "kids" came back and couldn't stop talking about how expensive things were. They said they even split a hamburger, because it cost five dollars. They piled the Honda Civic with their stuff and headed back to St. Louis. It was up to us to store the gifts. They didn't have room for them in their apartment.

On September 10, Momma had her eightieth birthday. She was tickled and talked a lot about making it to eighty. We didn't really celebrate. She always got bunches of cards and would put them up on the mantel in the living room. Kim and Toren had given her their gift before they left. I can't remember what I gave her. It's like a fog now, but I remember she was happy.

The wedding had caused Momma to change her Optimist Club meeting, always on the first Saturday of the month, to Thursday, a luncheon meeting. That meant I wouldn't be able to help her serve, though I can't say I was disappointed about that. Momma did all her usual cooking: homemade rolls, lemon meringue pie, chicken salad, fresh fruit tray, and an assorted hot vegetable tray. She always insisted that everything was presented with the proper garnish. When Momma came to our house for dinner, she would often bring parsley, because I never remembered to buy it. This became a standing joke. Anyway, she entertained the ladies, and they had a great time. After it was over, Momma told me everyone wanted to know about Kim and said they missed her as the hostess. We chuckled. I said, "I guess they didn't miss me doing the dishes." Momma laughed and said *she* sure did, because she was so tired.

Before we knew it, it was time to go to St. Louis for the reception to be given by Toren's parents. The Joneses had sent out formal invitations, so this was an opportunity for our family to get to meet their family and friends. We were all looking forward to this trip, even Daddy not complaining, because this was for Kim. So Gene, Momma, Daddy, and I left early

Friday and would return on Sunday.

When you travel with Daddy, you leave at the crack of dawn. Now, this certainly wasn't Gene's cup of tea. The later you leave the better, as far as he's concerned. In fact, whenever we go anywhere, I mean anywhere, it takes him forever to leave the house. He goes back in at least twice while I sit in the car drumming my fingers.

Momma and I have this ritual. As soon as we hit the city limit, we always begin snacking. This really irritated Daddy, but we'd just laugh and keep eating. Gene and I switched off driving, and we stopped a couple of time at rest areas, so both Mom and Dad did fine.

We pulled up in front of the Marriott around three o'clock St. Louis time. Momma and Gene got out to go register. She was so excited that she looked at me and said, "How should I pay for this, with cash or my Visa?" It was as if she had never been in a hotel before, and here they had been all over the world. I said, "Momma, just put it on your Visa and save your cash." She smiled and followed Gene into the lobby.

When Gene made the reservation, he booked two rooms on the concierge floor. The rooms had a spectacular view of the Arch, downtown, and the Mississippi River. There was a lounge next to our room with complimentary hors d'oeuvres in late afternoon and continental breakfast in the morning. Momma just loved this. She got plates for herself and Daddy and took them to their room. She said, "How much is all this going to cost? No wonder you all don't have any money, if you travel like this!" Both of us chuckled. I said, "Oh Momma, you're a mess. You and Daddy have stayed in plenty of hotels." Her reply was, "Yes, but it's been so long since we've been anywhere." I knew this was true, and it made me feel good about having them with us.

As soon as we settled, I called Kim. She lived only ten minutes from the hotel. It seemed I had no more than hung up the phone when Kim bounced into our room as excited as ever. She started talking nonstop: Where do you want to go for dinner, do you want to go out to the house, do you want to see our apartment, and on and on. She finally calmed down. After resting a little, she took us to their apartment, then to the new house. I mean the foundation. The model home was closed, so we said we'd come back the next day.

By this time, we were all really tired and ready to eat. Kim was still so thrilled to have us there that she was hyper. She rattled off three or four places we could go to eat, but I suggested we just go someplace close since I was so tired. I could tell Daddy was tired, too, but Momma was hanging in there and enjoying every minute. After going to a couple of places that were too crowded, we ended up at Union Station.

Saturday morning, Kim took us on a whirlwind tour of St. Louis. That is, Gene, Momma, Kim, and myself. Daddy was smart. He was tired and refused to leave the hotel room, saying he'd be fine.

We headed back out to the new house and were able to go through the model that was like Kim and Toren's house. Momma was so excited. She and Kim were already planning how Momma and Daddy could come up and spend a week, and Momma would make all the window treatments. While Gene took pictures of each room, Momma was busy checking out everything, saying over and over how much she liked the floor plan.

Ready to move on, Gene decided he wanted to go down to Soulard Market, all the way back downtown. He was looking for some pears, to make preserves, which meant walking through this large open air market. Now, I zipped up and down the wide aisles, because I was too tired to stand around and go slow. But Gene and Momma were busy comparing prices on things and commenting on all the fruits and vegetables, so I had to stop and wait for them.

When Gene, Kim, and Momma caught up with me, Gene and Kim had this weird look on their faces. Then they started laughing. Gene said, "Carol, come over here. I want to show you something." I walked over to where they were, and Gene and Kim burst out laughing. I had no idea what was going on. Momma didn't seem to know either. Well, I'd been standing right in front of all these wooden cages full of chickens. They both knew better than to make me aware of it, because if I had turned around and seen them, I would have had a heart attack on the spot. I'm terrified of anything with feathers and was shaken by the very idea of standing there. So the joke was on me. But that did it. I said, "Let's go."

Kim said she needed something from the shopping center, about 15 miles from downtown, and of course, Gene and Momma said OK. So again I was out-voted and off we went. Momma and I were in the back seat and Momma was taking in everything while Kim was busy narrating. Since I'm not a shopper like Momma, Gene, and Kim are, once we got inside I was dragging along behind them through Lord & Taylor's. Kim picked up some stockings, while Gene headed for the men's department. We finally left there and went out into the mall. When we got to my favorite shop, where they make fresh pretzels, we stopped, sat down on a bench, and enjoyed pretzels and soda.

Suddenly, Kim said she wanted to go to Dillard's to check

on her gift registry. Now Dillard's was at the other end of the mall, so I said, "I'll wait for you right here." Then she turned, saying, "Come on, Momma!" "Kim," I said, "Momma is tired." But Momma said she'd go. So off they went, though I could see that Momma *was* tired.

When we returned to the hotel, Daddy was in good spirits even though we'd been gone for hours. Kim went home, and the rest of us took a nap. Later, Kim and Toren came by the hotel, and we followed them to the reception. The affair was being held at The Vagabond's Club House, which was a huge old mansion on a beautiful, tree-lined boulevard.

Luke and Norma had really put this affair together. We had a receiving line in the gigantic foyer. Guests mingled throughout the house, from the big living room to the oak-paneled library, then on to the dining room where a huge table was beautifully set with food. At least 150 people came, bearing gifts and best wishes. Kim and Toren were just as bubbly as always, and it was obvious that Toren's family really liked Kim. That made me feel good. I told several of his family members, including Luke and Norma, that I was counting on them to take good care of my "baby."

We had a wonderful evening, and even Daddy was sitting over on the couch talking to various people, telling them about not being able to see. Momma sat and talked for a long time with Kim's good friend Kristi. Kim's friend Fay was there and gave us all hugs and kisses. Kim had been Fay's coach when her son Joshua was born. So although we were strangers, yet we weren't, because we knew of so many of their St. Louis friends. Norma and Luke were surrounded by their family and close friends just as we had been at the wedding. Everyone bent over backwards to make us feel welcome and to make sure we enjoyed our stay. And, of course, we did.

When we returned to the hotel that night, we all crashed. Kim and Toren came by the hotel, and we all had breakfast in the coffee shop. We checked out and followed them out to his parents house, arriving about noon. Norma had insisted we come by before we left town, and she had prepared food, so we ate. Then several of her friends and family stopped by. This turned out to be the most relaxed time I had the entire weekend. One of Toren's cousins, Alvin, got into a long discussion with Daddy. This kept him from worrying about getting on the road. Gene and Luke were downstairs, and the rest of us just sat around talking. It was just plain old family, and in that moment, I felt the union of our two families. In my mind, on that Sunday afternoon, September 18, 1994, the Jones, Bouldin, and Walker families became one.

twelve

It was three o'clock before we left the Joneses' and got on the expressway. Gene started out driving, and Daddy was asleep before we got out of St. Louis. I suddenly didn't feel well. Now, I had felt fine when I got in the car, so when I say sudden, I do mean sudden. The farther we drove, the worse I began to feel. I started having chills, so I took some Tylenol. Everyone was quiet, and by now, both Momma and Daddy were asleep. The further we went, the worse I felt. I covered up with my travel beach towel, had my pillow, and put back the front seat. After a while, I fell off to sleep.

Gene ended up driving the entire trip, and by the time we got home, I was really sick. The only good thing was that both Momma and Daddy had slept most of the way home, but I knew they were exhausted. When we dropped them off, Daddy was glad to be home. Momma told me to go home and get in the bed.

That night I was sick as a dog: chills, fever, aching, and I couldn't go to sleep. Poor Gene was exhausted and was out like a light. The next morning I had a bad case of something. I called work, took some medicine, drank my coffee, and went back to bed. Momma called later, saying she knew I wasn't going to be able to go to work and that I should go to the doctor. I told her I'd probably be much better after I got rested. Well, the following day, here comes Momma with her homemade chicken soup. Of course it was good, but by this time, I was really weak. Momma insisted I go to the doctor, so I made an appointment for the following day.

The doctor said I had pneumonia. She gave me several prescriptions and sent me home to bed. During the second week, I still wasn't getting better and went back to the doctor, Gene taking me. They took another chest x-ray and more blood tests. My doctor said I was dehydrated and that I could either go across the street to the hospital emergency room and get an IV, or be admitted overnight. I said, I'd go over to emergency. Then I was to go home, continue the medication, and stay in bed.

Momma came out to the house almost every day, bringing me first one thing, then another. She also checked on Aunt Minnie for me and did her laundry. Aunt Minnie called me about every three days, cracking jokes. She really was concerned, and that's her way of showing that she's worried. Gene is always a real Trojan. He'd come home, fix food, change the bed, and was his usual upbeat self. I was so sick!

This whole ordeal lasted almost three weeks. It was really a letdown after having such a wonderful celebration and trip to St. Louis.

thirteen

The bout with pneumonia took a lot out of me. The doctor said I was just run down and in a weakened state. But I had to get back to work and to my regular routine. Although I talked to Momma almost every evening, I only went by their house maybe twice a week. Usually on Sunday, if I didn't go by after church, they would come out to our house. Then at least once or twice during the week I'd stop by for our kitchen table discussions.

Now, every Tuesday evening, Momma would go to her good friend Pinkie's house for their Tuesday night sewing class. Pinkie—Mrs. Bailey—is a seamstress who makes all kinds of elaborate window treatments, bedspreads, pillows, etc., for local interior decorators. Tuesday evening was her night to socialize with Momma, Aunt Cora (not a relative), Ms. Mooney, and Mrs. Thomas. I have no idea how many years these ladies had their Tuesday sessions. Rain, sleet, or snow didn't stop them.

They worked on individual projects, like quilts, clothing, draperies, and made items for Momma's Christmas Bazaar, or just helped Mrs. Bailey, I guess. I never went to one of the sessions, but I gathered that they each shared news and gossip. This day was the highlight of the week for Momma. She usually got home around eleven o'clock, or a little after. When she came through the garage door, she would have more material scraps from Mrs. Bailey. If the weather was bad, Mrs. Bailey would call to make sure she got home safely. The other ladies all lived on Mrs. Bailey's street, but Momma had at least a twenty-minute drive.

On Monday, I stopped by the house for my usual kitchen table conversation. Thinking back to that day, nothing was the least bit unusual. Wednesday evening Daddy called and said Momma was sick. She had been sick since Tuesday, throwing up, and couldn't eat anything. He didn't want Momma to know he had called me, so I told him I'd call back a little later.

When I hung up the phone, I immediately thought, That's just like Momma not to let me know she was sick—I bet she caught something from me.

When I called her, I knew instantly that she really felt bad. But as usual, she thought she'd feel better in the morning. I told her as soon as I got to work I'd make her an appointment for the doctor. She didn't even put up a fuss.

Thursday, I left work at noon, picked up Momma, and took her to the doctor's. She was weak, and I had to help her get in and out of the car. She was also very quiet, which is unusual. Since Momma is a very proud, independent, private lady, I let her go back to the examining room with the nurse. I stayed in the lobby. After fifteen or twenty minutes, the nurse came to get me. When I walked into the examining room, there sat Momma, looking so frail.

When the doctor came in, she had the nurse take Momma down the hall for a urinalysis. Momma could barely walk. The doctor turned to me and said, "Your mother is dehydrated and jaundiced. I'm going to admit her to the hospital today so we can run some tests and get some fluids into her. Her weakened state is probably because she hasn't been able to keep any food down since Tuesday."

Now, quite honestly, the idea of her going to the hospital didn't upset me, but the word *jaundiced* was freaking me out. I hadn't noticed Momma being jaundiced. This immediately sent my mind sailing back to 1981 when my mother's younger sister, Ruth, went into the hospital because she was jaundiced. Aunt Ruthie died of what I think was cancer of the liver within five months.

I told the doctor I was sure that Momma would be upset about going to the hospital. I also told her that I had just gotten over pneumonia and that Momma had been around me a lot. She didn't acknowledge what I said. She just went on to say she wanted to get her started on an IV as soon as possible. She'd notify admitting and talk to Momma when she came back. She was aware that Momma was the care provider for Daddy and that this had been difficult for her. She asked if Daddy's care would be a problem for the family, and I told her no. I would see that Daddy was taken care of. The primary concern was to get Momma back on her feet. I told her I wanted my cheery Momma back. She smiled and said, "Yes, you're right. She's a lovely lady."

Sure enough, when Momma came back into the room, I looked at her face and it had this yellow glow. She was definitely jaundiced, and I hadn't noticed at all. The doctor told Momma all she had told me and explained that the dehydration was probably why she was so weak. Momma looked up at her with tears in her eyes and said softly, "I don't want to go to the hospital."

I was really trying to hold myself together. "Now,

Momma, you don't have a choice. You're always taking care of other people, and now you need some fluids so you can get your strength back."

She said, "I know, but what about Daddy?"

I told her I'd take care of Daddy and that he'd be just fine.

Momma was pretty quiet as we drove home from the doctor's office. I knew she was upset about going to the hospital, but she also knew how bad she felt. I talked to her about catching something from me. I don't know whether she bought that or not, but I believed she had. What seemed to bother her most was that she couldn't eat or keep anything down. Finally, I got my nerve up and mentioned the jaundice. I knew Momma's mind had gone to the same place mine had when she heard the word. I told her I hadn't noticed it until the doctor mentioned it. But that now I could see a yellow glow to her face. Playing doctor, I said that this was probably because she'd been throwing up bile. Momma said nothing. She probably wanted to tell me that she was sick, not stupid.

fourteen

When we arrived back at the house, I pulled up out front and helped Momma out of the car. She walked slowly across the front walk, wearing her black and white tweed coat and her distinctive black hat. She was really tired and sat down at the kitchen table. Daddy was in his usual chair. I told him that Momma was going to have to go to the hospital. He was shocked. He immediately said, "I'll be fine. Just take care of your mother. Help her get her things." Momma just sat there saying nothing. She wouldn't eat and Daddy wouldn't either.

I helped Momma put a few things in her bag, called my house and left a message for Gene, put something out for Daddy to eat, and then we left for the hospital. On the way, I told Momma, "I'm glad you're going to the hospital because you can get on an antibiotic and get the fluids you need. That will knock this virus out in a couple of days."

Weakly, she said, "Yes, I guess you're right."

By the time we arrived at the hospital, it was four o'clock. Once again I parked behind the hospital, in the emergency lot, after dropping Momma off at the door. When I got to the door, Momma was sitting right inside. We walked over to admitting. Here was the same lady who admitted me about two weeks ago, when I had to have the IV. She immediately got Momma a wheelchair, processed her admission, and took us up to her room. She was so pleasant and kind to us. She said all kinds of reassuring things to Momma. She could tell that Momma was not only really sick, but that she was very depressed about being in the hospital. Although, I thought it was more fear than depression.

Exhausted, Momma finally got in bed. The nurse came in and introduced herself. They did the initial chart stuff, like temperature and blood pressure. Within minutes, Momma drifted off to sleep. I called Daddy, and he was sitting right by the phone. I told him Momma was settled and asleep. Now, he hadn't eaten a thing, so I fussed at him about that. He just kept saying, "Take care of your mother." Then he went into detail about how he could fix his cereal in the morning and scramble an egg. Mind you, Daddy couldn't (and wouldn't) try to cook anything. Momma cooked a full dinner every day or had leftovers. We called her "the queen of leftovers," and I mean there could be a tablespoon left. She didn't believe in throwing any food away and didn't. She fussed at me all the time for wasting food and cooking too much. I swear I don't like leftovers, but I could pop in, sit down at their table, and enjoy all the various dabs of this and that. It tasted great, I guess because it was Momma's cooking.

I told Daddy I was going to stay at the hospital for a while and that I'd call him when I got home. I also told him to call Marsha. He said he didn't want to, didn't know what to tell her. So I said I'd call her. On second thought, I didn't want him getting her all upset.

When I got off the phone, Momma was snoring. I pulled the cover up, because she had been worried about being cold in the hospital. She looked so pale, and I could definitely see the yellow tinge on her face. I sat there convincing myself that she had the flu. The nurse came back to put in the IV. Momma didn't like needles, so what can you expect but that the nurse had a hard time finding a good vein. After two attempts, she called in another nurse who got it in on the first try. I could then see Momma relax. When they left, Momma said for me to go on home, that she knew I was tired. But I told her I'd stay awhile and for her to get some rest. Again, she drifted off to sleep.

I left the room and went downstairs and out in front of the hospital to smoke. I was so tense, and the cool evening air felt good. I watched all the cars and people coming and going. I'd tell myself to calm down, but my mind was racing. After all, Momma was never sick!

I went back upstairs, and since Momma was still sleeping, I called Gene and filled him in. He was very reassuring. I told him I'd be home a little later. Momma was still snoring, so I got up and looked out the window. We were on the seventh floor and the view was to the northwest, across the city. It was twilight. Beautiful. You could see the Mill Creek Valley and all the way over to the west side of town. I can't remember what I was thinking, but I do remember the view. It was peaceful and calming.

Before I left, I told Momma I had a meeting in the morning and that I would call her after that. I again assured her that I would take care of Daddy. Back home, I called Marsha, Kim, and Daddy, then headed straight to bed. As a matter of fact, I was already in the bed when I made the calls. I was wiped out. What a day.

The next morning I had a conference at 8:30 on the other side of town. I arrived a few minutes early and sat in my car trying to decide whether or not to call Momma. I decided it was too early and went on into the building. Of all days to have a difficult case conference. I was truly preoccupied. It seemed that the talk went on and on. Finally, I abruptly said, "We must bring closure to this meeting because I have another commitment." I'm usually willing to let everyone vent and have their say, but not on that morning.

I couldn't get to my car fast enough. I called Momma from the school parking lot. She sounded weak and somewhat disoriented. She told me they were going to operate and that the surgeon would be in around noon. I said, "Operate, operate on what?" Momma mumbled that she wasn't sure. They had taken her down early for tests, and her doctor had been in early, too. "I'll be right there," I said. It was 10:45.

Pulling out of the parking lot, I called my secretary and told her what was happening and that I wouldn't be in the rest of the day. She was very sympathetic. I then called Daddy and found out Momma had called him. He wasn't clear what was going on either. He was also upset, so I had to reassure him that I'd find out what was going on and let him know. I said I'd come over later and pick him up so he could come see Momma. He told me he had eaten some cereal and for me to "Just take care of Momma."

When I arrived at the hospital, I went directly to the nurses' station and asked for Momma's nurse. They paged her. When she came up, I told her who I was and that I wanted to know what was going on with my mother, that she seemed confused and had mentioned something about surgery. The nurse told me that Momma's doctor had been in early that morning and that she had called in a surgeon to consult. "As a matter of fact," she said, "I think he's still here working on her chart."

As we spoke, he walked out of a little room off the nurses' station, and the nurse called him over. I introduced myself and asked what he knew about my mother's condition. He said the tests they had done indicated she had acute gallstones. That's not the medical term. He had scheduled surgery for the next day, Saturday, at 10:00 a.m. He assured me that, although she was in a weakened condition, they felt she could tolerate the surgery. It needed to be done immediately. I discussed my mother and her role as the care provider and main support of our family. I also shared with him that although Momma had undergone surgery for Meniere's disease, ear surgery, about ten years ago, she hadn't been sick and always had regular checkups. Then I asked the tough question: "Why the jaundice?" He drew a picture and said the gallbladder was enlarged, full of stones and pushing on the liver, which could account for the jaundice. He indicated that when he actually went in he'd be able to tell more about how involved things were, but he didn't anticipate any problems.

Now, I had never seen this man in my life, but I made no bones about telling him, "This is my Momma, and I want to know everything there is to know and I want her to get excellent care." He looked at me and said something reassuring. Despite my upset, I felt good about this man. He was very patient with me and yet very direct. I respect that. He made it clear that I could feel free to call him at any time if I had questions, gave me his card, and promised he would not leave the hospital without talking to me after the surgery. Well, you can't ask for more than that.

I went into Momma's room, and she looked up and smiled. I sat on the side of the bed and told her I had just met with the doctor. I told her everything. She said they punched and probed on her that morning and that it hurt so bad. She said that he told her about surgery but she couldn't remember everything. I spent a long time talking to her about what the doctor had said, and although she wasn't pleased, she seemed to relax a little. I knew Momma trusted me. I think the fact that I seemed to understand what was going on made a difference to her. I also reminded her that she had taken care of me when I had my gallbladder removed. She smiled.

I called Gene and asked him to go by and pick up Daddy so I could stay at the hospital. He also needed to stop and get Daddy his favorite, a Wendy's and a Frosty. Momma didn't want me to stay, but I told her I wasn't going back to work. I said, "Gene is going to bring Daddy over after work, so I don't have anything to do but keep you company." Momma smiled and said she knew I was tired. "I am, but you take a nap, and I'll take a nap over here in the reclining chair."

Before I drifted off to sleep, I suddenly remembered Aunt Minnie. I hadn't talked to her or been to see her since Tuesday. She didn't even know Momma was sick. I remember thinking, I'll go by this evening, when I leave here.

Later, I went outside and smoked a couple of cigarettes, went into the gift shop and bought some snacks and a soda, then went back upstairs. Momma wasn't eating much of anything and was on a liquid diet. She hated the broth, saying it was too salty. At least she was getting a supplement through the IV. She drifted in and out of sleep, snoring loudly. While she was awake, I did most of the talking, and she seemed fine, not disoriented like on the phone that morning. Momma told me not to get Marsha all upset. I said I'd call her after she got home from work and let her know about the surgery. Momma was adamant that Marsha didn't need to come to

Cincinnati, and I agreed with her. Within the next hour, Marsha called the hospital and talked to both Momma and myself. This was good for both of them.

Daddy and Gene came in. Daddy went over and gave Momma a kiss on the cheek. She got on him about fixing himself something to eat, and he was busy telling her not to worry about him. Gene gave Momma a pep talk, and she just smiled and smiled. He assured her that things were going to be fine and that he would be praying for her. This pleased Momma.

That evening when I finally got home, I crashed. I didn't have to go by the nursing home because Aunt Minnie had talked to Daddy, so she knew what was going on. Daddy, in his abrupt manner, had told her not to worry Carol, saying I had enough on me right now. I reminded myself to make sure I let her know about how the surgery went, because she was just sitting there worrying about Harriet. I knew that.

I talked to Marsha a long time that night and told her I'd keep her posted about the surgery. The doctor told me it would last at least three hours. I would go over before they took Momma down. When I hung up the phone, that was all she wrote. I was asleep within minutes. It was 10:30 and had been a hard day.

Just after midnight, I was awakened by voices. I heard "Momma, Momma!" I turned over, and there stood Kim and Gene. Kim had left St. Louis after work and driven home. I responded just like Momma, saying, "What are you doing here?!" She said she wanted to come home and help me with Momma and Daddy. By then, I was awake, so we talked a short while. Kim got really upset, then calmed down. Finally, we all went to bed. We had to get to the hospital early, before they took Momma to surgery.

fifteen

Kim and I arrived at the hospital around nine o'clock Saturday morning. When we walked into Momma's room, her face lit up.

"Kim, what are you doing here?"

Kim went over and gave Momma a kiss and said, "Oh, I just thought I'd come home and make sure Momma took good care of you."

Momma smiled and chuckled. "Kim, you'll never change. You just don't want to miss anything. You've always wanted to be in two places at the same time." We all laughed because that's certainly the truth.

Kim and I were standing at the foot of the bed when Momma said, "You know I had this dream last night about Ruthie. She was in her back yard, down by the fence. I was calling her to come and see the new washing machine. She just looked at me and didn't say anything. I saw her plain as day. You know they say, if the dead talk back to you in your dreams, then they're coming to get you. But Ruthie didn't answer me."

There was total silence. Then I said, "You know, Momma, if someone was coming to get me, I can't think of a better person than Aunt Ruthie."

Momma smiled, and that ended the conversation. She

didn't seem bothered by the dream or by my comment. Kim was standing there like a statue, in shock. I was also in shock, not believing what I had just said. Here was Momma, going into surgery within the next thirty minutes, telling us very calmly and matter-of-factly that she had this vivid dream about her dead sister. I can't put into words my feelings at that moment, and I doubt if Kim could either. Later, Kim said, "Mom, how could you say that about Aunt Ruthie?" I told her I didn't know, it just came out. I was shocked myself.

An emergency heart surgery delayed Momma going downstairs. It was close to eleven when they came up to get her. She made us promise that as soon as she got back to her room we would see that she got her teeth put back in. Momma wouldn't let anyone see her without her teeth. This was a big thing, and we always laughed, but she was very serious. Of course we promised we would and for her not to worry about that.

Kim and I went downstairs with Momma and waited with her about 30 minutes before they actually took her back to surgery. Momma was scared, but calm. The nurse told us to wait in the surgical waiting room and that her doctor would meet us there following surgery. When we got to the waiting room, it was already 11:30, so I said to Kim that we ought to go out and get something to eat.

We left the hospital, and I drove downtown to this restaurant where you can get giant corned beef sandwiches. Kim went in, picked them up, and we drove back to the hospital parking lot. While we sat there eating, Kim said again, "Momma, how could you say what you did to Momma when she told us about her dream?" I told her, again, that I really didn't know, that it just came out, and that was all I could say about it! We talked about how scary it was for Momma to have this dream when she was so sick and undergoing surgery. We talked and talked, I guess trying to assure each other that everything was going to be fine. I must admit, it was a real lift to have Kim there, since Gene was at work.

When we went back to the waiting room, it was almost one o'clock. I opened the door, and there sat cousin Lorna with a cane, her foot in a cast. Her eyes were as big as saucers when she looked up at us. Immediately, she said, "Where have you been? I've had them paging you. I've been up to Aunt Harriet's room. I've been down to the cafeteria. They told me you were here in the hospital."

"What's wrong?" I demanded, afraid, because Lorna was so agitated.

She realized she had scared us and quickly said, "Nothing. I just couldn't find you!"

When I told her we left the hospital to get corned beef sandwiches, she flipped. Lorna is always comical, so she got all dramatic telling us that she had everyone in the hospital looking for us. We all laughed.

Lorna was like another daughter to Momma. She practically grew up at our house and is the same age as Marsha. The same thing happened the next generation, Lorna's daughter, Heather, growing up with Kim, like sisters. So Lorna coming to the hospital meant a lot to us. It helped the time pass more quickly, too, because Lorna keeps you in stitches telling her stories. Daddy calls her "giggle box."

I called Daddy several times, just to let him know we were still waiting. He was sitting in the kitchen, right beside the phone. I hated to think how tough this was on him, but he couldn't tolerate sitting at the hospital. He needed to go to the bathroom frequently, and he had difficulty functioning in unfamiliar surroundings.

It was 2:25 when the door opened and in walked the surgeon's nurse. She said the doctor had sent her out to let us know that the surgery was going OK, but it would probably be another two hours. I asked what was wrong. She said the gallbladder was loaded with stones and that there were more stones in the duct. The doctor had removed the gallbladder and was trying to get into the duct, but it was twisted. She said, "This is delicate work, and that's why the additional time." But the bottom line was that Momma was tolerating the surgery well. As the nurse turned to leave the room, I said, in front of Lorna and Kim, "Tell him to check her liver good!" She called back, "I will."

This made us all a little more relaxed. I thought it was nice of the doctor to send his nurse out to tell us the surgery was going to last longer. At least we didn't have to sit there not knowing why it was taking so long. We had an explanation. I retold the story about when Gene was at this same hospital for back surgery. The doctor told me it would take four hours. After five hours, I got up and went through the double doors that were boldly marked DO NOT ENTER! I walked all the way down this wide hallway, to where I saw Gene's bed. Just as I approached the bed, outside of the surgery room, here came what seemed like an army of nurses, wearing their green scrub suits. I guess they weren't used to crazy ladies coming back into the surgery unit. "No, I can read," I said, "But I've been sitting in that waiting room alone and no one has bothered to tell me anything." They quickly escorted me to the door, saying Gene was doing fine, but that it was taking longer than expected. Lorna, Kim, and I cracked up.

I called Daddy again to tell him the latest. He told me Aunt Minnie had called twice. I had to remind him that people would be calling and not to be rude or abrupt with them, that people were just showing their concern. Just say we don't know anything yet and thank them. This all goes back to Daddy's change in temperament, since his vision problems. When he's upset or agitated, he comes across very short with people. Most people don't understand and are offended. Momma dealt with these mood swings twenty-four hours a day. Sometimes she ignored him, and sometimes she let it get to her.

In about an hour, here came the doctor. He called me into a small conference room. I introduced him to Kim and Lorna. He said he had removed the gallbladder, which was greatly enlarged and full of stones. As a matter of fact, he said her gallbladder was "the size of a pound roll of sausage." It should have been the size of a walnut. He went on to say that there were still several stones left in the duct, but the area was so inflamed he had to stop the operation. If he had continued, there was a possibility of rupturing the duct, which could be fatal. He said, "The difference between a good surgeon and a surgeon is knowing when to stop, even if the surgery isn't complete."

He was hopeful that, when the inflammation cleared up, the stones in the duct would pass naturally. If not, then he would refer Momma to a gastroenterologist. That would mean she would have to come back in a couple of weeks as an outpatient, and they would run "the scope" down to try and dislodge the stones. But the main thing was that none of this could be done until the inflammation was gone.

Now, the doctor went into great detail explaining

Momma's condition. I asked him directly, "What about Momma's liver?" He said, "I examined her liver, and it was clean." We all looked at each other and breathed a sigh of relief. It was like a terrible cloud had been lifted from this whole ordeal. However, he seemed astounded that Momma's gallbladder was so enlarged and said it was pushing on her liver and other organs. He expected, though, that her liver function tests would improve now, with the removal of her gallbladder. All in all, he felt Momma had tolerated the surgery well for an eighty-year-old lady.

He mentioned that gallbladder disease tends to run in families, and Lorna and I started rattling off family members who had had their gallbladders removed: Momma's mother, sisters, Ruth, Edith, and maybe Clara, brother John, and Lorna and myself, and others might have problems also. Then Lorna and I looked at Kim and said, "Oh, my! You're probably next!" Kim said, "Not me!" We all laughed. Then, on a serious note, the doctor said, "Well, if anyone in your family has gallstones, they should consider the surgery and not wait for problems. Your family history most certainly justifies the procedure."

Lorna left to go home. Kim and I made all the phone calls and went back upstairs to wait for Momma to come out of the recovery room. By then we were giggling and letting out all the tenseness of the previous few days. Things sounded good. The doctor was very encouraging, and our prayers had been answered. Momma was OK.

Momma didn't get back to her room from recovery until nearly six. The very first thing I did was help her get her teeth in her mouth. As groggy and miserable as she was, she managed to help. She looked frail and pasty white and was obviously in pain. She was on a machine where she could control the morphine dosage by pushing a button. I guess you'd say she was semiconscious. She responded when you asked her something, but drifted in and out. I was concerned because she was in pain, moaning, but couldn't seem to understand that she had to push this button for the pain medication. I kept telling her, but she would drop the button out of her hand and didn't seem to understand. Kim and I would take her hand and help her push the button. Then she'd get the medication and doze off for a while. When the nurse came in, I told her this seemed to be a problem. She pinned the button to Momma's gown, but that didn't work either. Momma was just so out of it that she didn't understand.

Now, after I had surgery and was on this machine, Gene said, "You came up from the recovery room, still out, and were pushing the button every few minutes." He had to ask if I could get too much medication. The answer is no; it's regulated to prevent that. But you get better relief than from shots, which can wear off before the next one is scheduled.

Gene and Daddy came to the hospital. Momma was sleeping on and off, but she was aware we were there. She smiled when I told her that the doctor said everything went well. I told her the worst part was over and that evoked a groan. I guess I'd said another stupid thing. After all, here she was suffering. Kim was so helpful, sponging off Momma's face and again helping her push the button. Daddy sat in a chair close to the bed. He kept trying to talk to Momma, but she was out of it. If she answered, it was so softly that he couldn't hear her. He has no hearing on the left side, so he kept saying, "I can't hear her. What's she saying?"

Daddy was upset. He wants no part of hospitals and

doctors. And I think he felt helpless, too, not in control. This, after all, was his wife of over fifty years, who was never sick, and he depended on her for everything since his vision problem. Here he was unable to do anything to comfort her. Or that's what I thought, just watching and being a part of the dynamics of this situation.

Daddy went home with Gene, Kim and I deciding to stay on awhile longer. When we finally left the room that night, we stopped at the nurses' station. I told Momma's nurse for that shift that I was very concerned about her not being able to push the button for the pain medication. I asked that they watch her because she wasn't getting enough medication to keep her comfortable. She was very nice, but as I would later find out, she didn't pay any attention to my concern.

Leaving the hospital was hard for both of us that night, but we were totally, and I do mean totally, exhausted. As a matter of fact, I didn't have any business behind the wheel of a car for twenty miles. We stopped and got some fast food, then went home and straight to bed. Oh, I called Marsha again, and that was it. Kim, too, was out like a light.

Our intention was to get to the hospital early the next morning, but that didn't happen. We were both so out of it that it was 10:30 before we even got out of the house. The phone had been ringing since early that morning. Although I was appreciative of people's concern, it really slowed us down. Finally, I told Gene that he'd have to deal with the phone.

When Kim and I left the house, I told her we'd better run by the nursing home to see Aunt Minnie. That way we could just pop in and not stay long because she'd understand that we were on our way to the hospital. After all, I hadn't been there since Tuesday. I knew that not only was she worried about

Momma, but that I needed to get her laundry.

Aunt Minnie started crying as soon as she saw us. We told her that Momma was going to be just fine. She had told probably everyone in the place, and of course they knew Momma from her visits. Within a few minutes, she was fine and went back to laughing and making sure that she got some attention from both of us. She had her dirty clothes ready, and I put them in the yellow plastic bag. When she walked us to the front door, I told her that she could call me in the evenings. I had to remind her that I now had to check on Daddy, get him food, take him to the hospital, check on Momma, as well as keep up with her and the nursing home. She asked then about my job, if I was going to have to take off, and I told her that I'd be off at least the coming week, until I got everyone settled. This seemed to satisfy her. Despite her desire for all the attention she could get, she was able to express concern for my getting some rest, too. That was really sweet.

By the time we arrived at the hospital, it was close to noon. When we walked into Momma's room, she was awake. She looked so frail and sad, but when she looked up at us, there was that smile. Her voice was soft and weak. She told us they had gotten her up that morning and how tough that was, also that she'd had a terrible night. She was in such pain, and the nurses didn't give her anything. She said that when she pressed the nurse's call button they would ask what she wanted, then wouldn't come in to check on her. Momma was still on the machine. I asked her if she was pushing the button to get her pain medication. Although she was much more alert, she still seemed confused about using the button. She did say she tried, but that the button would get lost in the bed.

She was afraid of getting too much medication, too. I went through explaining this again and assured Momma that the medication was controlled and would keep her comfortable.

I was already functioning like a space cadet. I was exhausted, both physically and mentally. Now, after hearing Momma tell us how she suffered in pain all night and that she didn't tell her doctor when she came in early that morning, I was on the verge of exploding. I left Kim with Momma, went out to the nurses' station, and asked for Momma's nurse. When she came to the desk, I asked what kind of report she got from the night shift nurse. She began telling me that Momma had a "restless night" and wasn't taking her pain medication. She also said that she had put on her call light several times, but when someone would go in, she was asleep. They weren't able to hear her when she pushed the call button, because her voice was so weak.

Now, I must say, the nurse was nice, but I was getting more agitated. Still, I kept my cool and asked what the surgeon's orders and her doctor's orders were from that morning. She pulled her chart and said her vital signs were improved and that there was no notation of any change in medication. The nurses were to get her up and sit her in the chair for as long as she could tolerate this.

Having been through surgery several times myself, I know that when you're in the hospital, if you don't speak up, then you're going to be overlooked. They have limited staff and often respond only when someone asks for help or questions something. I politely thanked her for the information and told her that I would be calling the doctors to get Momma off the medication machine. She needed to be on scheduled pain medication, because she was confused and unable to handle this method of self-medication.

I went to the pay phone in the lounge, called both doctors, left messages for them to contact the nurses' station to change the orders, and left the phone number of Momma's room as well as my home number. I fully explained to each office the reason for my request. To my surprise, within a couple of hours, in came the nurse to disconnect the machine. She then gave Momma some medication that went directly into the IV. Soon Momma dosed off and the moaning and discomfort seemed to subside.

Gene brought Daddy over later and saw to it that he got dinner. Daddy seemed pleased that Momma was more alert when she was awake. He sat patiently in the chair, holding his cane, as we all passed the day and evening. Daddy would get up and toddle over to the bathroom every hour or so. We all knew he didn't have any business using the patient's bathroom, but we ignored this. The visitor's bathroom was at the other end of the hallway. When he had to go, it was right then, so there was no sense trying to get him all the way down the hall. We had enough to deal with without adding that to the list.

Although this medication change caused Momma to sleep more, at least I knew she was getting some pain relief every four hours.

sixteen

Those days in the hospital were emotionally tough on Momma. She was frightened at night and talked about the noises outside her room. Her room was across from the main laundry chute for that floor and large bins of laundry were constantly being dumped, which made a lot of noise. We tried to explain this to Momma. I even walked her across the hall and showed her the laundry chute. She seemed to understand, but would say that at night the noise was different. She felt she wasn't safe and that there was a party going on at the other end of the corridor. Even walking in the hallway, Momma was disturbed about being there. She wanted to come home.

It was Thursday before she was released to come home. Momma's doctor had called me and said that the pathology report indicated there was "a cancer contained in the gallbladder." My heart skipped as I continued to listen. "If this had been diagnosed prior to the surgery, then the recommended procedure would have been to remove the gallbladder." Since that had just been done, she didn't anticipate any further concern. The surrounding tissue samples were "clean," and the only concern now was getting the remaining stones out of the duct. Again, this would be done in about three weeks, provided the inflamed area was clear. I remember distinctly questioning about "the cancer." She assured me it was contained "within the gallbladder." What a relief! What a gigantic relief! I was truly thankful! She had told Momma all of this that morning.

Although Momma was weak as could be, she was finally more like herself because she was going home. We had taken most of the flowers home, so when I arrived to pick up Momma, all she had was her bag. The nurses had gotten her dressed, and she was sitting in the chair waiting for me. When I entered the room, out popped that big smile, and she started shuffling to get out of the chair. I told her to wait, because the nurse had to sign her out and bring a wheelchair. I think it must have seemed like an eternity to her, but it was really only about twenty minutes.

Momma was talking again, softly and slowly, but talking. As we drove down Clifton Avenue past the elementary school, Momma turned to me and said, "My doctor said I had cancer."

Without hesitating, I said, "Yes! She told me you had a cancer which was contained in your gallbladder and that removal of the gallbladder took care of it. Is that what she told you?"

Momma very faintly said, "Yes."

I went on to tell her how we had asked the surgeon about her liver and that he had told Lorna, Kim, and myself that it was "clean." I explained about her having to undergo the "scope" in a couple of weeks, to get the remaining stones. She definitely wasn't looking forward to that, but Momma had always been a real trooper. I was sure she'd handle it OK, especially with time to get past this hospital experience.

After this brief but intense conversation, Momma switched to talking about Marsha coming to spend the week with her. Anytime you mentioned Marsha, Momma always said, "Don't worry Marsha!" But even though Momma had talked to her every day from the hospital, she couldn't wait to see her, and I knew this. Not only was Momma glad to see her, but so were the rest of us. Kim had already gone back to St. Louis, so Marsha decided to come when Momma got home. This was a tremendous help to me and gave her time to care for Momma and Daddy. This, I'm sure, meant a lot to her and made her not feel so removed from all that was going on.

Daddy was thrilled to have them both home. Of course, he had missed all the attention he got from Momma and still couldn't absorb the fact that she was so weak. Not only was she weak, but Momma had really lost weight. I hadn't realized, but she had been losing weight for months. That was why she was able to wear her suit from the Orient at Kim's wedding. I was aware now that she was extremely frail. I felt stupid that I hadn't been aware of it before the surgery. It's the same as my not being able to look at her and tell she was jaundiced. I can't begin to explain how and why I could have overlooked this and not seen signs of her slowing down. Thinking back, there were definite signs. She wasn't cooking as much and would pick up fast food for their dinners. She would say she was tired and didn't feel like doing or going somewhere. She talked about not having the energy to get things done around the house. She said things didn't taste good to her anymore and she didn't have an appetite. Now, all these things meant something, and I hadn't stopped long enough to put them together. In my mind, Momma was the same as always, and I truly didn't focus on her slowing down.

I can't begin to tell you how much I think about these things. I feel I was really remiss in not recognizing sooner that Momma needed to go to the doctor. And here, just a few weeks before this surgery, she had been coming out to my house every day, bringing me homemade soup and worrying about me getting better, when she didn't feel well. We had walked Momma all over the mall in St. Louis, and she hung in there, not missing one store. All of this makes me feel guilty and just plain stupid and selfish.

During the week that Marsha was here, various members of the family came by the house to visit with Momma. While our cousins Paul and Iris were there, Marsha asked Iris if she could take Momma to the surgeon the next week. Iris graciously agreed, which meant I didn't have to take more time off work. When Paul's mother and father were ill, Aunt Edith and Uncle Clayton, Momma was a regular visitor. She checked on both of them and dutifully stood by each of them until their deaths. Aunt Edith was Momma's oldest sister. Again, that's why Marsha and I called Momma Florence Nightingale.

It was the first week in November when Marsha returned to Maryland. We all really missed her when she left, but she called Momma almost every evening.

seventeen

Marsha really helped get things back on track. She contacted the Board of Elections and let them know that Momma would be unable to work at the polls on Election Day. She arranged to give Momma's monthly symphony ticket to her friend, Mrs. Lemon. She let her bowling and golf teams know that she would be out for a while. She contacted the president of The Optimist Club, Mrs. Kelly, who came over to get the books and bank statements since Momma was the treasurer. She contacted the Wilberforce Alumni Club because, again, Momma was treasurer. Also, Momma was chairperson for Circle Six at our church, Mt. Zion United Methodist. And on and on. This was a big relief to Momma, because she was very diligent about her duties. This gave her less to worry about.

During this period when Momma was first home, Aunt Minnie would call her in the evening from the nurses' station. Being able to talk to Harriet herself made Aunt Minnie feel a part of what was going on. Even though I was back on my regular schedule visiting her in the nursing home, she took great pleasure in letting me know she had talked to Momma. Believe me, she gave everyone reports,.because staff members there often asked me about Momma.

Each week, Momma received three visits from an RN and three from a home health aide also. These services were certainly helpful to us, and Momma was greatly appreciative of the care. I sent a letter to her doctor on November 7 and told her I was concerned because Momma didn't have an appetite and was losing weight. I also let her know that she was terribly weak and needed help with everything. She responded, and an appointment was scheduled for Momma to come into the office on November 18.

On Wednesday, November 9, Momma had her first post-operation appointment with the surgeon, cousin Iris taking her. The surgeon told both Momma and Iris the same thing her doctor had told me, that there was "cancer contained inside the gallbladder" and, again, that if this had been diagnosed prior to the surgery, the remedy would have been to remove the gallbladder. His main concern was the remaining gallstones. He referred Momma to a gastroenterologist who would do the procedure that I called "the scope" (ERCP) to remove the stones. Momma wasn't the least bit pleased about this, however she understood.

When I went by after work that day, Momma was exhausted from the trip to the doctor. Her voice was faint, and she couldn't eat her dinner. She wasn't hungry, and the food didn't taste right. She didn't want me to fix her anything else to eat. She just wanted to rest. Momma told me that even though she was tired, it had felt good to get out of the house. She said Iris had been so kind, helping her in and out of the car.

This was the longest period Momma had been stuck in the house. She couldn't stand more than a couple of days without going somewhere. And, of course, she was used to going to the grocery and Brentwood Shopping Plaza almost every day. But she felt so weak she didn't complain.

During this time, I turned the screws on Daddy. He hadn't been to a medical doctor in several years. He didn't believe in doctors and could always tell you a million reasons why he didn't need a checkup. Well, I came down on him hard, saying that if anything happened to him, he needed to have a doctor in order to get into the hospital. I told him that Momma got in immediately because she had a doctor. Don't ask me if he believed this, but I did get him to see a doctor on November 11.

When I called the doctor's office to make the appointment, I insisted that I also talk directly to the doctor. I introduced myself and began to tell him about my Dad and my concerns. I made it clear that he had better do whatever needed to be done when I brought Dad in on Friday, because there were no guarantees I could get him back for a second appointment. He needed a good physical. The doctor seemed very understanding. It's terrible, but I also told him that if Daddy didn't like him, then he could forget it. We both laughed!

Momma couldn't believe I had pulled off getting Daddy to go to the doctor. She was pleased and encouraged him, while he went on about it being a waste of money. Not only did Daddy go, but he went back for a second appointment. Fortunately, the tests came back OK, and of course he got the opportunity to say, "I told you so." But, in our minds, it was worth it. After all, Daddy would turn 84 on the ninth of December.

Months earlier, Gene and I had made reservations for the National Alliance of Black School Educators Convention to be held in Los Angeles the week before Thanksgiving. I was worried about going, but Momma wanted us to. Of course, this was just like her. Kim arranged her schedule so that she could come home and stay while we were away. This certainly was a relief, but I called Momma every day during the five days we were gone. She always said she was doing better. She enjoyed having Kim around, so that lifted her spirits. I guess we had a good trip, but I remember how worried I was and how eager I was to get back home. Gene felt I needed the trip to relax and unwind. That was easier said than done, but at least I didn't have to worry about Aunt Minnie. She just wanted me to bring back something for her.

It had been a month since the surgery, and Momma was still not showing any improvement. She was scheduled for the scope on November 21 at 6:45 in the morning. This was to be an outpatient procedure. I arrived around 5:30 and found that Momma had managed to get herself dressed. She was pale and exhausted as we walked slowly to the car. It was dark and cold, and all I could think was that she didn't need to be out. I was upset, because I knew she was scared, but I told her that when they got these remaining stones out she would begin to feel better. Momma was very quiet and just said, "I hope so."

Fortunately, when we got to the hospital, they were ready for her. The procedure would take about forty minutes, and then she'd be in recovery for an hour. Well, it took much longer, and when I went back into recovery, I found out that Momma had had some internal bleeding. The doctor said he was unable to get all the gallstones out of the duct. He had removed several, but there were two more lodged deep in the crinkled duct neck, and he had to stop due to the internal bleeding. He felt Momma couldn't tolerate more time. This meant he would have to do this again. My heart went all the way down to my feet. I knew how upset Momma would be. This is not a pleasant procedure, and she would have to come back again.

When I had this "scope" done years before, I called the doctor a son of a bitch. That wasn't my usual manner, but he was pushing this tube down my throat and I was gagging. He acted irritated and told me to stop gagging and be still. That's when I called him this. Even with the tube in my mouth, my words were clear enough for him to understand. The nurse said, "Just relax now. Just relax." So I knew first-hand what they were doing to my Momma.

They decided to keep Momma overnight for observation. This was really going to trip Momma out and also upset Daddy. However, as I had told the doctor, I wanted nothing but the best care for her. If she needed to stay, then I wanted her to stay, especially since there was concern about the internal bleeding. They would give her a blood transfusion and IV for dehydration. I told him she would be upset and that I would talk to her.

I knew by this time that Daddy was probably a wreck. After all, he was sitting at home not knowing what was going on. I called around noon, and he answered on the second ring. He was in his usual spot, in his kitchen chair. I told him everything as calmly as I could. His voice was shaky, and of course he wanted Momma to stay if there was something wrong. I told him I would be over later that afternoon to bring him to the hospital. His main concern was that I stay with Momma.

Before I went back to the recovery room, I called Gene and told him what was going on. He said he would pick up Daddy after work, get him something to eat, and bring him over to the hospital. That was a tremendous help.

When I went back into the recovery room, Momma was

more alert and was cold. I got another blanket for her and sat on the side of the bed. I told her what the doctor had said. She looked at me and shook her head. She said faintly, "I don't want to stay here again." I told her that they were going to give her a blood transfusion and an IV. This would help her get her strength back. I explained how she had lost some blood during the procedure and that they wanted to make sure there was no further bleeding. Since Momma has never been a stubborn, hard-headed person like me, she reluctantly agreed that this was best.

As soon as she got to her room, they started the IV and blood transfusion. She drifted off to sleep. Later, I helped her out of bed so she could go to the bathroom. Momma was so, so weak! This was exhausting for her, so when she got back in bed, it wasn't long before she was asleep again.

Later, Daddy and Gene came over. Momma was alert and wanted to know what time I'd pick her up the next day. I told her to call me in the morning or I would call her. I said, "As soon as they say you can go home, I'll be here." She smiled, and we all said good night. It was about 7:30.

When I got home, I called Marsha. She sounded terrible. She had a bad cold. Of course she was upset about Momma and would call her after we talked. It was Monday evening, Thanksgiving just three days away. Marsha had wanted to come home again for the holiday, but not sick. No way did Momma need to catch anything. We agreed that if she didn't feel much better she should stay home and get some rest.

Momma was released about noon on Tuesday. She was still weak, but her spirits were a little better because she was going home. On the way home, I drove her to Brentwood Plaza, and she sat in the car while I went in the drugstore to pick up some more Ensure. Since she wasn't eating, I had been buying Ensure, and she tried to get down at least a can a day. This was hard, but she tried.

Marsha didn't come for Thanksgiving. She was still sick. Since I was in an unknown state of exhaustion, I ordered the turkey dinner special from Kroger's. Gene and I went over to Momma and Daddy's, and I set the dining room table, made Daddy some corn pudding, and the four of us sat down to our store-bought turkey dinner. Momma sat there in her robe and moved the food around with her fork. She ate very little and then went back to bed. The weaker she got, the softer her voice got. This was upsetting to Daddy, because he couldn't hear her.

eighteen

As the days passed, Momma still didn't seem to be gaining her strength back. She said food tasted salty. She really tried to eat, but just couldn't. I kept going to the grocery and bringing in all kinds of things: juices, fruits, raisin bread, cheeses, whatever I saw that I thought she might eat. Nothing really worked, and Momma kept fussing at me about spending money on all this food. I really don't mean fussing, but she actually didn't even want to see food. But I couldn't help but keep trying to hit on something she'd be able to eat.

Some days, I'd drive across town to the grocery before I went to work and pick up something I had thought of the night before. I'd take it by Momma's and then go to work. Or I would go by during my lunch hour and fix soup or something. She never refused but could only eat a few spoonfuls.

Most days Momma got dressed, went in the kitchen, fixed Daddy's bowl of cereal, tried to eat something herself, then went back to bed. Fortunately, their house was one floor, so she didn't have to climb any steps. However, just going from the bedroom to the kitchen became more and more of a chore. By now, I had called and talked to her doctor three or four times about her being so weak and unable to eat. She continued to say it was going to take time, and I had no reason to doubt that. After my own gallbladder surgery, in my late forties, I was off work about eight weeks. And here Momma was eighty.

With all this going on, I still managed to go to work, even though I missed plenty of days. When I took the job with the county in October, 1993, I told them I'd need to be able to take time off without pay, but I had no idea I would need to be off so much. I had all the responsibility for Momma, Daddy, and Aunt Minnie, plus keeping up with my job and things at home. There's no way I could have done all of this without Gene, either. He pitched in and did whatever was necessary to keep things together. He was there when I came home exhausted, both mentally and physically, every day. My car phone was put to good use, and my secretary, Marlene, helped me to keep on top of my messages. I'm proud to say that despite all that was going on and the time I missed at work, I didn't let any of my cases go unattended. This is quite an accomplishment since I was totally and completely exhausted.

Aunt Minnie was very concerned about Momma. She hadn't seen her since mid-October, and it was now early December. She called her a couple of times a week, but it seemed to go over her head that Momma wasn't getting better. At least that's what I thought. She was as demanding of me as ever and wanted her fair share of attention. I was still going by the nursing home, picking up her laundry, and taking whatever things she wanted.

By this time, she practically ran the nursing home. She was an official greeter for new residents, was on the resident council, and was participating in all the activities. She knew the names of every staff person on each shift and as much of their business as she could find out. She was no longer incontinent, and I have no idea what had happened with the dementia. It seemed to me she was back to her crazy self. However, they couldn't get her to attend any of the trips outside of the building, and I couldn't get her to come home with me. Even trying to get her to go visit Momma didn't work. She'd say simply, "I'm not able to" or just plain "No!"

Daddy was also in denial. He kept telling Momma that she shouldn't spend so much time in bed. He felt that was why she didn't have any strength. He also was on her about eating, which would aggravate Momma. Daddy was very upset about Momma being sick. I think he felt helpless, both because he was unable to take care of her, as well as needing so much help himself. He kept thinking of old home remedies from the dark ages and insisting that we try them. We got into an argument because he wanted us to make Momma what he called an "asfittity bag" to put around her neck. I don't even remember what this was supposed to accomplish, but both Marsha and I said No! Momma sat there too weak to say anything. She just smiled and shook her head. She knew Daddy meant well, but she wanted no part of his idea.

Kim would call from St. Louis and Kristy from Maryland. Both were concerned about their grandmother. Their calls were always a highlight for Momma. Momma was interested in hearing about the progress on Kim and Toren's house. They hoped to move in sometime mid-December. Since Momma had been to St. Louis and seen the model and their lot, she could follow along with Kim's almost daily reports. She and Kim talked about how she and Daddy would spend a week when Momma felt better, so she could make Kim's window treatments. This had been the plan since Kim and Toren started building the house. Now, it was more Kim talking about this than Momma, but she was as pleased as she could be about their new home.

Momma didn't have a lot of company, and this was good, because she was so tired. Her brother Earl came by a couple of times a week and would always bring her fresh fruit. He was so upset about Momma's illness. She was the one he depended on to help when he was sick. He lived alone and had been very ill off and on. He would call Momma in the middle of the night rather than call 911. Momma would get up and drive to his house, pick him up, take him to the hospital, and wait until he was admitted. He would religiously give Momma his wallet to keep for him. It seemed that Momma and her brother, Chuck, were the only ones who could hold Uncle Earl's money. Both of his sons lived on the east coast, so when he needed help, it was up to Harriet and Chuck. We would fuss at Momma for going out in the middle of the night, but if Uncle Earl called, she went. He has heart trouble and some kind of digestive problems. I was always telling Momma that he should call 911, because if he was having a heart attack, there wasn't anything she could do. Here we have an eighty-year-old lady taking her seventy-eight-year-old brother to the hospital in the middle of the night. His concern was that Momma should be there to keep his money and house key. But I must admit, Momma was the one who could calm him down and assure him that he was going to be OK.

Other members of the family dropped by, and Momma's friends either came by or called. She was in constant touch with her friend Pinkie and different church and club members.

Marsha talked to Momma almost every day, and I talked to Marsha probably every couple of days. She would come home for Christmas and was going to bring her whole family.

This gave Momma something to look forward to. But, in some ways, it made it a sad time for her. This was the time of year that she was usually bouncing around doing her Christmas shopping, decorating the house, sending out cards, and baking all her special goodies for the day-after-Christmas Family Open House.

nineteen

On Saturday December 3, I had to take Daddy back to the doctor for more tests and a blood pressure check at 10:15. He did fine. He was in a good mood but was nervous until the doctor told him everything had come back within normal limits. Even his pressure was fine. Since Daddy doesn't believe in medicine, this was great and he was proud to tell anyone that he had "No aches or pains." This was his standard line, and we all got tired of hearing it.

As soon as I got him home, I had to take Momma to her doctor for a one o'clock appointment. On her chart was written, "The patient comes in with symptoms of anorexia, early satiety, fatigue, burning epigastric pain at nighttime, and reflux symptoms in the middle of the night. Her weight has gone up only half a pound.... The patient still has two retained stones in the common bile duct. She is scheduled for a repeat ERCP on the 16th." The doctor started Momma on Zantac and wanted her to return in three weeks for more blood tests.

I can't begin to tell you the number of times I talked to Momma's doctor. She was always very reassuring and was patient with my questions and concerns. She always called me back, and if we were in the office, would see that we had private time to talk. Momma was becoming more and more withdrawn. I noticed that she was only answering the questions asked her. She herself didn't question anything. While we were in the office, it took every bit of her strength to get weighed, go to the bathroom for a urinalysis, come back to the examination room, and go through the physical exam. I sat there feeling so helpless. I knew Momma just wanted to be left alone. It was an emotional experience for me, but I had to remain upbeat and positive in front of Momma. There was no doubt how weak Momma was, and she looked as pale as a ghost.

On the way home, I did most of the talking, and she just sat there. Every once in a while she would acknowledge what I was saying, but not through conversation. She didn't have that much energy. But one thing for certain, she was dreading going through the ERCP again. Make no mistake about it, this was very upsetting to her, and as a result, I felt like shit. I was telling her all this Pollyanna stuff about if you just do this, then you'll feel better. Don't ask exactly what I said, because it was just whatever came to mind at the time. It was so upsetting, and I felt so helpless. I couldn't seem to do whatever had to be done to get Momma back on her feet, and this was tearing me up inside.

When we got back to the house, Daddy was waiting at the kitchen table for a report. Momma sat down, totally spent. She didn't want anything to eat but did take a couple sips of Ensure. I helped her get undressed and back in bed. Within minutes she was asleep.

Again I tried to talk to Daddy in the kitchen. He truly believed that if Momma just stayed up she would regain her strength. I kept trying to tell him that she couldn't, and he kept saying, "The bed will sap all your strength." This was almost a daily discussion, and I would get irritated. I did realize, however, that Daddy was as upset as I was, and he didn't know what to do any more than I did. We both wanted to make Momma bounce around like she always had, and this wasn't happening. Our ways of handling this were different, and I realize that my ideas were as lame as his.

In some ways, I was on a roll. I had managed to get Daddy referred to a podiatrist. His toes were terrible. He had overgrown nails and a nail fungus. This was something he had been treated for before, but of course he hadn't gone back to the doctor. Somehow with Momma being sick, he was a little more compliant. I got him an appointment for the 12th, and the doctor recommended an over-the-counter ointment that he should use daily. I stopped by the pharmacy and picked up the ointment, which cost over ten dollars. Like a dummy, I told Daddy how much it cost. He flipped his lid and wouldn't use the medication nightly, because it cost too much. Momma just said, "It's *his* toes, so it's up to him to use the medication."

Before I knew it, Friday, December 16, had arrived. I picked up Momma about 6:00 a.m., and we were again off to the hospital. There really was no change in her condition. She was still weak, pale, unable to eat, and talking less and less. She was definitely depressed about her recovery, and there was nothing I could seem to do. Momma dreaded going back for this test, so I tried to reassure her that this time they would get the two remaining stones in the duct—then she would begin to feel better. In my mind, I felt this had something to do with her loss of appetite. This wasn't like Momma not to eat, because she loved food and would eat almost anything. She always fussed at us about picking over food.

When we arrived at the hospital, again they took her

right away. I waited in the lounge for the doctor. This time, it took about an hour. When the doctor came out, he said he had removed the two remaining stones. There wasn't the bleeding that had occurred before, and Momma would be able to go home after she came out of recovery. What a relief that Momma wouldn't have to go through any more of these intrusive procedures! I couldn't wait to call Daddy and tell him that we'd be home in a couple of hours. I could hear the relief in his voice. I went to the recovery room to sit with Momma. She was wiped out, but she smiled when I told her they had gotten the two remaining stones. She looked up at me and said softly, "Can I go home?" I said "Yes, just get some rest first!" She smiled again and dozed off to sleep. Momma had really been through the mill.

As I sat in the chair close to her bedside, I watched her sleep. I began to cry, tears rolling down my face like water. I was so thankful that this ordeal was finally over, and I prayed to God that Momma would soon regain her strength. Everything from the last two months seemed to come down on me while I sat there watching her sleep. My mind was racing. I was talking to God. I was reassuring Momma and, I guess, reassuring myself.

Two nurses in the recovery room came over and asked if they could get me anything. I told them I would be fine, that I just had been so worried about my mother. One of the nurses said, "I remember when you and your mother were here last month. She seems to be a lovely person." I told them how Momma always took care of everyone else and that I was devastated seeing her so weak and ill. In their way, they were very kind and reassuring. Then I had to pull myself together, because I couldn't have Momma waking up and seeing my face the color of a fire engine. This would have made her believe something was terribly wrong. I have always said the Lord works in mysterious ways. On this day, in that recovery room, the Lord allowed me to release some of my burden and pulled me back together before Momma awakened.

twenty

December 20, I called the doctor again, telling her how concerned I was about Momma, that she wasn't getting better at all. The RN came twice a week and just kept saying she needed to eat more, that she needed more food and fluids. This was all well and good, but Momma couldn't eat. There was no doubt about it, she could not swallow food. She sipped some liquids, but very little. The doctor told me to bring her in on the 22nd.

By now, this was a major effort, because Momma was so very weak. She was down to 130 pounds from her normal weight of about 155. Her chart on December 22nd read: "The patient has been feeling down, fatigued. Has hardly been eating since she was last in ten days ago, despite the use of Zantac. Says she feels short of breath just walking to the kitchen from her bedroom. Occasionally vomits, usually only in the morning. Gets down some breakfast, but seldom eats during the rest of the day, other than Ensure at night.... Chest x-ray showed evidence of a probable right lower lobe pneumonia." They took the usual blood tests to check potassium level, liver profile, and who knows what else. Momma's doctor felt an antibiotic would clear up the pneumonia. Because of Christmas, it would probably be Monday before she got the labs results back.

Saturday was Christmas Eve, and Marsha was coming home with all her family. Momma was looking forward to seeing them, as always, but was so weak she could no longer get up and get dressed. She was also talking less and less. Her voice was very faint and weak, when she did speak. She had stopped talking on the phone and spent most of her time in the bed. It was a real effort for her to get up and walk across the hall to the bathroom. She was now on the antibiotic. I told Momma she would begin to feel better when the medicine kicked in and she was over the pneumonia. She didn't respond.

Momma had handled all the bills and business since Daddy lost his sight in 1989. The first of December, she sat at the kitchen table and wrote checks for the bills, her usually perfect handwriting noticeably shaky. Now, she asked me to go to the bank and get her some cash and money envelopes for the grandchildren. She also had me write some checks for her, and I got Daddy to sign them. This meant Momma was really sick. Never in all my years had Momma asked me to handle any of their business. Both she and Daddy were very secretive about their business. This was not a good sign, but Momma would get better. I knew she was going to bounce back. It just took time!

Saturday, December 24th, in comes Marsha, Kristy, Kevin, Michael, and Ronnie, Kristy's son. Marsha was totally shaken at seeing Momma and how much weight she had lost. She had no way of knowing how weak and fragile she was, because she hadn't seen her since the first week in November. There's no way to describe how upsetting this was, not only for Marsha, but for all the grandchildren. Kim would fly home on Christmas Day, in the afternoon. Toren had to work the 26th.

Kristy's son, Ronnie, was four years old and full of energy. He understood that Momma was sick, so he crawled up in the bed with her to show her his pictures and books. As sick as Momma was, this brought her great joy. This was her great-grandson, and he tried every way he could to cheer her. He had never seen Momma not up and about. He kept worrying about who was going to cook for us. Ronnie wanted to make sure that he had his Christmas dinner. In his little mind, that meant Momma was supposed to cook.

It was obvious that Kristy, Kevin, and Michael were pretty shaken also. They didn't know how to react, because their grandmother was not just their grandmother but their best friend. Momma wrote each of them while they were in college, sent them money, made clothing for them, and gave them advice. Not only did they love her, but they had the utmost respect for her. She had given of her love freely and without strings.

Marsha, Daddy, and I were dazed. I don't know how we continued to function. We had to pull together Christmas dinner, go over to see Aunt Minnie, and try to make the day as festive as possible. The traditional Walker Christmas started with breakfast at Momma's, then gift opening in the living room, and dinner would alternate between Momma's and her sister Willa's. Before Aunt Ruthie died in 1982, her family was part of this tradition. The only surviving member of her family now is our cousin Karen, and she lives in San Francisco.

But, on Christmas Day, 1994, we skipped the big family breakfast. Gene and I went over to the house around eleven. Momma was in bed but got up to come out to the living room. Needless to say, neither Marsha nor I had done much Christmas shopping. I had ordered a couple of things from a catalog and would give the kids money. Gene had picked up some toys for Ronnie. I recall Momma sitting in the big chair in the living room. It was Ronnie's job to pass out the presents. Although he couldn't read yet, he knew his name, his mother's, Kevin's, Michael's, and Marsha's. Momma's handmade felt stockings were hanging on the fireplace. As usual, Santa had filled each grandchild's stocking with candy, an orange, and an apple.

As Ronnie gave Momma a present, she would smile and say thank you. She sat there with gifts in her lap and others on the floor. Until Ronnie said, "Momma open your presents," I didn't realize that she was just sitting there. Momma didn't have the energy to open her gifts. Ronnie pitched in excitedly to help her. He ripped paper as fast as he could, then would give the gift to Momma. She would say thank you, but without her usual enthusiasm. I had bought her a new black purse for everyday. She just placed it slowly on the floor. Momma just wanted to get back in the bed. Daddy did his usual: getting trash bags for all the wrapping paper and boxes, always concerned about cleaning up the mess.

Christmas Day, we had dinner for just our immediate family. Aunt Willa's family didn't come over as had been the tradition. Marsha and I put together a nice meal, set the dining room table with the Christmas tablecloth, china, and, of course, a Christmas centerpiece. Gene had bought Momma a poinsettia early in the month. Marsha and the kids had put up the tree on Christmas Eve. So as always, Momma's decorations were up, and we didn't even forget her little arrangements for the bathrooms. Believe it or not, we did a pretty good job.

Kim's plane was due around 1:30, so Gene went to the airport to pick her up. When Kim came in, she lit up the house. A month since she had seen Momma, she, too, was shaken to see her in bed, so pale and thin and weak. When Momma looked up and saw Kim, she broke out in a smile. Kim sat on the side of the bed and began telling Momma about their new house. They had moved in on December 16th. Although Momma wasn't talking much, you could see that she enjoyed being surrounded by her family. After all, her entire life had been devoted to each of us, as well as to countless others. That was *our* place now. It was our turn to care for her.

Later that afternoon, we went to the nursing home to see Aunt Minnie. She started crying when we came in, but that didn't last but a minute. She was too busy opening her gifts and running her mouth. We had to hear about all the things that had gone on that day at the nursing home. Aunt Minnie was like a child experiencing Christmas for the first time. She was so happy. Earlier in the week, she had made it clear to me that she wouldn't come and have dinner with us. Even with Momma sick, she still wouldn't leave the nursing home. She hadn't seen Momma since the middle of October. But this didn't mean she didn't care. She cared deeply in her own way, just didn't want to visit. One thing I must say, Aunt Minnie was always very optimistic about Momma getting better. If I told her I was worried, she always came back with something very positive and supportive, then would crack a joke.

I think what Aunt Minnie enjoyed most about our visit was the opportunity to parade her family around, introducing us to everyone. My dear Aunt Minnie loved to brag. She would have those people believing we were in line for the English throne. This is what made her happy.

twenty-one

Monday, December 26th, was the day that Momma's family always came over for her "Holiday Open House." Everyone knew that Momma was sick, so the only people we were expecting were her sisters, Florence and Geneva, along with their families, from Columbus. Now, this was still a lot of people, but Marsha had pulled together food for them.

I had planned on getting over to Momma's early but couldn't seem to get myself together. I know it was just exhaustion, but that didn't count because I had to help. Minutes before leaving the house, the phone rang. It was Momma's doctor. She had gotten last week's test results back from the lab that morning, and Momma's liver profile was abnormal. She wanted Momma to go to the hospital on Tuesday for a CT scan of the abdomen, an outpatient procedure. My heart was up in my mouth.

The doctor kept saying she needed to get further information from this test. I, again, for the hundredth time, told her how weak Momma was and that she wasn't eating much at all. She acknowledged what I said, then went on to tell me to come by her office that afternoon to get the medical orders for the test the next day. When I hung up the phone, I went to pieces. It was as if I'd been struck by lightning. All along, I had continued to believe that Momma just needed more time to recover because she was eighty years old. Now they were telling me that her liver tests were abnormal. I had asked the surgeon to check her liver when she was operated on in October. He told, not just me, but also Lorna and Kim, that "her liver was clean." How could this be? How could this be?

By this time, I was crying, sobbing, my mind racing from one thing to another. How was I going to tell Momma that she had to go back to the hospital for another test the next day? How was I going to tell Daddy and Marsha? How was Momma going to be able even to get up and go to the hospital? How was I going to tell them what the doctor had said, when we all knew that Aunt Ruthie had died from cancer of the liver? How was I going to face Momma? I had been telling her everything was going to be fine and to be patient.

My entire body was just broken. Never before, and I do mean never before, had I even thought of Momma not getting better. There was no way this could be happening, not to my Momma. Kim and Gene had already gone. I walked from room to room crying, screaming NO!, praying, and telling myself to pull myself together. My world was coming down around me, and I felt totally helpless. I was used to being in control of things, and this was something I couldn't do anything about. I had always taken great pride in trying to fix things for others. I couldn't fix this for my Momma.

My whole career in education, as an administrator, was problem-solving, so that some child could get appropriate services. In other words, my daily work for the last thirty years had been resolving issues, solving problems, helping others. Now, here I was, totally wiped out, because I couldn't help Momma.

After what seemed an eternity, I got in the car and drove

over to Momma's. As I came down the street, I saw the cars from Columbus. The doctor's office was further west than Momma's, so I decided to talk to Marsha and Daddy first, then I'd go get the orders for the CT scan. I didn't have to say much. My face told the story that things weren't good. Everyone started hugging me when I came in, and I was glad to see them. My cousin Jackie was there from Hawaii with her new husband, Ron. This was the first time I had met him. The only time I saw Jackie was at Christmas, usually the highlight of the day for me. After Jackie married, she had moved from Los Angeles to Hawaii. Jackie is a Registered Nurse, along with her sisters, Donna and Susan. They were all there. I realize now that they knew how sick Momma was.

I told Marsha about the doctor's call and that I was going to her office now. We didn't tell Daddy, because there were too many people around. And I didn't go into the bedroom to see Momma, because I looked like I'd been hit by a Mack truck. Poor Marsha couldn't even go to pieces, because she was right there in the middle of all this commotion. I left quietly and went to the doctor's office. I didn't get to see her, but the nurse had the paper ready. She looked at me and must have seen my sadness, because she said something very kind. I wished her Happy Holidays and left.

By the time I got back, there were more people there. My cousins, Gerald and Eric, had brought their father, Uncle Chuck, over to see Momma. Uncle Chuck, Momma's older brother, was very close to Momma. Momma always said that Chuck had worked to help her through college, sending her money for her tuition and spending money. Uncle Chuck had been diagnosed with Parkinson's Disease and was very emotional. When he saw Momma, he started crying. This seemed to be part of his illness, because he also cried when he was happy. Yet, I knew these were not tears of joy. His little sister was very sick, and this upset him greatly. Both Gerald and Eric were pretty shaken also. They both were close to Momma, along with their brother Michael. Uncle Chuck's wife, Aunt Florence Ann, was also ill and unable to come out. However, she called Momma regularly and was very concerned about her.

This day that was usually so warm and festive was quiet, somber, and subdued. Momma stayed in the bed. Family was in the kitchen, living room, dining room, and family room. A couple of people were always in with Momma, but she wasn't talking much. She just didn't have the energy. Yet, I do believe she was glad to have everyone in her home. This was the way it was supposed to be the day after Christmas.

Finally, I pulled myself together enough to go in and talk to Momma. We were alone, and I sat on the bed and told her that I had talked to her doctor. I told her about the tests coming back abnormal and that Marsha and I would take her to the hospital the next day for a CT scan. I didn't say that it was to be of her abdomen, and she didn't ask. I did tell her that it was like an X-ray and wouldn't hurt. She listened and didn't visibly react. I made it clear that this was an outpatient test and that she didn't have to stay at the hospital. I felt like shit! I had to keep myself together, and I knew this. Momma was so tired that she acknowledged what I had told her without saying a word. Then she just drifted off to sleep.

I went in the kitchen, pulled out the doctor's orders, and showed them to Jackie, Donna, and Susan. They each tried to be encouraging and not jump to conclusions, but it was obvious that they knew this was serious. Their Aunt Harriet was really sick. Momma was as close to a second mother to Jackie, Jeannette, Donna, and Susan as you could get. Aunt Florence, Momma's younger sister, and Uncle Jean had lived with us during the early years of their marriage, between his military assignments. Both Jackie and Jeannette were born here, so there had always been this bond. It's the same with every other member of the family. Momma had some kind of special tie with each of them, and she was the one they all called on for support.

I don't know whether it is a strength or a weakness, but I've always been one for detail. I think, talk, and write in detail. Since this is my way, I told my cousins everything that had gone on since October. I told them about the number of times I had called Momma's doctor about her being weak, not eating, losing weight, depressed, and so on. I told them about how all four doctors had said the same thing: "It's just going to take time" and for her to try to eat more. Yes, there were four doctors Momma had seen. Her regular internist, her partner, oh, two partners, the surgeon, and the gastroenterologist, that makes five doctors. They sat there letting me unload, but said very little. And now the doctor tells me that her liver profile came back abnormal. What's going on?

I went in the bathroom and washed my face. I just had to pull myself together. I'm sure I looked like a wild woman. By this time, Gene had arrived. That meant I had to tell all this again to him. As I could expect, he was reassuring and told me to calm down, to stop jumping to conclusions. I agreed, but deep down inside, I knew something was terribly wrong.

We didn't tell Daddy until everyone had gone. Then we just said that the doctor was concerned about Momma's test results and wanted to do a CT scan the next day. Neither Marsha nor I spoke of Momma's liver, yet I would bet anything that both our minds were in the same deep pit. The kids, Kim, Kristy, Kevin, and Michael, were concerned, but their minds hadn't taken a leap to something more serious. And little Ronnie was just enjoying himself and looking in on Momma every few minutes.

If I had to stand before a judge, I would swear I didn't get any sleep that night. At least it didn't seem that I did, and I was tired beyond words. My insides were churning, and my mind raced from one thing to another. But, mostly, I worried that there was something wrong with Momma's liver. I took a valium, and it didn't do a thing. I kept telling myself that things were going to be OK. Most of the night, I prayed for Momma and for strength. It was one long night, and I was not looking forward to taking Momma back to the hospital.

The next morning, I went over to get Momma. Marsha and I helped her get dressed while she sat on the side of the bed. This was pure torture, because she was so weak. There's no way she needed to be going out in the December cold for this test. It was just asking for her to catch something, but we had no choice.

We bundled her up, and Marsha and I walked her to my car. At least the car was still warm. When we got to the hospital, I pulled up to the emergency room door. Marsha got out and helped Momma inside, while I parked the car. When I walked into the waiting room, Momma was already in a wheelchair. We went over to admitting, and here was the same kind lady who had helped Momma, Daddy, and myself before. She smiled and took us right away. We answered all the questions for Momma as she just sat there.

Fortunately, we didn't have to wait. They were ready for Momma. The technician came in the waiting room and brought three cups full of pink barium for Momma to drink. Marsha and I looked at each other. We knew Momma couldn't drink all that stuff. Momma turned up her nose and tried to take a swallow. She softly said, "I can't drink this." We told her just to try. It was making her sick to her stomach. It was making me sick, too. I felt like we were being mean, trying to get her to drink this stuff.

Finally, I got up and went back to the technician, who was sitting in a little office. I called to him and told him there was no way my mother could drink all that barium, that she hadn't been able to eat or drink anything for some time, and that it was making her sick to her stomach. The man said he'd see if she had enough in her so it would show on the film. I went back and told Momma she didn't have to drink any more. I guess she drank almost a glassful.

Marsha and I talked while Momma got the CT scan. It was Tuesday, and she had planned to go home on Saturday. She was concerned and upset about how weak Momma was and felt she needed more care. Since the home health aide only came three days a week, for about three hours a day, Momma was still trying to get up and do things. Daddy was no help and needed someone to take care of him. Marsha said she would take sick leave and stay a little longer.

I knew she was right about Momma needing more care. I also knew that Momma was getting worse, not better. I had been off work so much that I had decided to work over the holidays, except for the 26th. Now here it was the 27th, and I was still off. Marsha was trying to convince me that if she stayed longer I wouldn't have to take off. Finally, I said we could decide on this later.

When Momma was finished with the test, the technician wheeled her back into the waiting room. She looked like a little dried-up, pale lady. She was exhausted. I went to get the car, and Marsha helped her out of the chair and into the front seat. I had to pull out her seat belt and buckle it around her. She was too tired to talk but did say that the test didn't hurt. Again we were out in the cold December air, taking Momma home.

When we got her in the house, she complained of an upset stomach from the barium. She couldn't eat anything, nor would she drink any juice or Ensure. She just wanted to get undressed and in the bed. Momma slept most of the afternoon, snoring loudly. She now needed help to get to the bathroom. I don't remember that she had much to eat that evening. Momma had become so frail, so very frail.

When I left to go home that evening, there was no way I could go by the nursing home. I was so tired I don't know how I was even still standing. Thank goodness for Kim. She had spent the day at Momma's, so she drove me home and we picked up some fast food. We talked and talked, and she just kept telling me not to worry, that Momma was going to be alright.

Kim was the first and only grandchild for five years. She also lived here, so Momma and Daddy were her heart. I looked at her and smiled. Here was my beautiful, grown daughter telling her mother not to worry. Kim has always been a positive person, and even this difficult time wasn't going to get her down. She truly believed, and that was comforting.

twenty-two

As soon as we got home, I headed straight to bed. It was early evening, but all I could think of was getting some sleep. I was so tired. So very tired. In fact, I was so tired that I couldn't go to sleep. I couldn't get comfortable, and I couldn't get my mind off Momma. I finally turned on the television and propped myself up on the pillows.

Within a few minutes, the phone rang. I figured it was Toren calling Kim, but it was early. I reached over and answered the phone. It was Momma's doctor. Suddenly I felt hot all over. I sat up on the side of the bed when she said she had the results of the CT scan. I could hear her children playing in the background as she told me that the scan showed a large mass on the liver. I was totally shocked. "No, that can't be true!" The surgeon had told me after the surgery in October that Momma's liver was "clean." This was December 27th. How could this be? She said she was surprised, too, but this could definitely be the reason for her lack of appetite, loss of weight, and weakness. By this time, I was choked up but managed to ask what could be done.

She told me she wanted Momma admitted the next day, Wednesday, in the afternoon. On Thursday, they would perform a liver biopsy to determine if the mass was malignant. The results of the CT scan didn't look good, especially coupled with her symptoms. She couldn't say how they would proceed until she got the biopsy results.

We talked at length, and I asked all kinds of questions, but I don't remember what they were. I was floored. This was becoming more and more a nightmare, and I didn't think I could take any more. Momma's doctor was very tolerant and not only listened to me, but expressed her concern for our family, especially Daddy. She knew that Momma was the care provider for Daddy. She tactfully let me know we could no longer depend on her for his care. She assured me she would be in touch as soon as she had the results from the biopsy.

I hung up the phone and went to pieces. Gene and Kim came running into the bedroom, and I lost it. I kept saying over and over, "How am I going to tell Momma she has to go back to the hospital? How am I going to tell her that she has a mass on her liver? How am I going to tell Marsha and Daddy? How am I going to tell Momma that she has to have a liver biopsy?"

Gene was pacing the floor, and Kim and I were crying. I felt like I was going to lose my mind. I just couldn't do this! This was my mother! How could I be the one to tell her this, when she's already depressed about being sick. After all, I'm the one who had been telling her for two months that she just needed to take it easy. I am the one who kept reminding her that there was nothing wrong with her liver, because that's what the surgeon told me. How am I going to face Momma? She hates the hospital and is afraid there at night. What am I going to do?

As I rambled on and on, Gene and Kim hugged me and told me to pull myself together. Gene kept saying, "You have to be strong for your mother and father." Kim stopped crying and began to say, "I know Momma's going to be alright. I know it!" The problem was that, suddenly, I didn't.

I had to call Marsha. What was I going to say? We both knew this would upset Momma terribly. I called and told Marsha to get Daddy on the phone. I told them both that the doctor had just called and that the CT scan had showed a mass on Momma's liver. They both gasped. I continued, saying that Momma had to go back to the hospital the next day for a liver biopsy on Thursday to determine if the mass was malignant. By this time, I could no longer keep my composure. I began crying as I talked. Marsha, too, was crying and choked up. Daddy was trying to sound brave and told us, "We must put this in the hands of the Lord." I agreed, and of course, so did Marsha. We talked further, then Daddy hung up. Marsha and I talked on about how frightened Momma was of the hospital. We both knew this was going to be terrible.

Don't think for a minute that I hadn't been praying all along, but on this night, I cried out to God to help us all. I was so frightened! I was in such pain and didn't want my mother to have to go back to the hospital again. Yet, I knew I wanted them to do whatever it took to make her better. I have a complete blank about how Momma was told about going back to the hospital. But I do remember going over the next morning and helping her dress and pack her bag. Her youngest sister, Willa, was there. Momma was sitting on the side of the bed when she said to Marsha and me, "You get my things together. I don't want Willa to go through my things and see that I'm not a good housekeeper." We all laughed, but Momma was serious.

Now, everyone in the family knew that Aunt Willa and their late sister Ruthie were immaculate housekeepers, but

Momma's house was by no means dirty, junky, or untidy. As a matter of fact, she did an excellent job, considering all she had to do. She complained to me every day about having to sweep up Daddy's tobacco from his pipe, off the kitchen floor. Since he couldn't see well, his corner of the kitchen was always a mess. I would tell Momma not to let it get to her. After all, it wasn't as if he dropped tobacco all over the house; he stayed in the same spot with his pipe. This never satisfied Momma. She hated the mess and would say that this is what she did all day, clean up tobacco.

Aunt Willa was a nervous wreck and walked from room to room. Her career had been that of a medical secretary in the Pathology Department at what was now University Hospital. She knew every disease, symptoms and prognosis. She had said things all along about being upset about Harriet. I knew she had discussed Momma's symptoms with the doctors at work. However, I guess we hear what we want to hear. So she knew that this new-found mass on the liver was bad news. She wasn't saying much, but I knew from her face and nervousness that she was upset.

That afternoon Marsha and I again made the trip to the hospital with Momma. Once again the same lady was on the afternoon shift in the admitting office. She took one look at us and evidently knew that we were all distraught. Momma's doctor had already called, and she knew that Momma was being admitted. She went through the necessary paperwork quickly. When she asked Momma if she had a Living Will, I answered, "No!" She asked another lady to take over her station while she took us up to the room. This was very kind. She turned and said to Marsha and me, "Let me know how your mother gets along."

Again we helped Momma get undressed and into her gown. She was already wiped out and was asleep before the floor nurse came to check her in and take her vitals. They started an IV, which had become more difficult because Momma's veins had collapsed. After two times, she called in another nurse. This really hurt Momma, and all Marsha and I could do was stand there. It hurts deeply to see someone going through this and to know there's nothing you can do.

We stayed at the hospital pretty late that evening because Momma was restless. By this time, Marsha had decided to stay for another week. The kids, Kristy, Kevin, Michael, and Ronnie, would leave Monday and drive her car back to Maryland. Marsha would stay the next week and fly home the following weekend. This meant that when Momma got out of the hospital, Marsha would be there to take care of her, and I could go to work.

Along with everything going on at home, my office was moving to a new location during this holiday week. My plan had been to be there, but fortunately, I had packed up my office and labeled things before the holiday, so all the movers had to do was to put my furniture, files, and boxes in my new office. Thinking back, I don't know how I had gotten all this done, but I was thankful it was one less thing to worry about. I had even brought home all of my plants. And, as terrible as it might seem, I had completely forgotten about Aunt Minnie. Kim was still home, so I told her to go over, pick up her laundry, and cheer her up.

The following day when we got to the hospital, Momma was almost delirious. She was weakly moaning and looked up at us saying how bad they had hurt her. She had been taken down early that morning for the liver biopsy. I had no idea they would puncture her abdomen and probe inside to get tissue samples and not put her to sleep. Whatever anesthesia they gave Momma was evidently local, and the test was terribly painful. She couldn't stop saying how they had poked and prodded in her stomach. She had tears in her eyes, and Marsha and I were just crushed. Momma was never one to complain, so this liver biopsy must have been a bitch. It had hurt my Momma badly. She begged us not to let them put her through any more of these tests. This was just terrible.

I remember getting up on the bed and telling her that it was all over now and she just needed to get some rest. Momma was slipping before our very eyes. Her color was pasty. Once again there was that yellow tinge to her skin, indicating jaundice. Even though at times she seemed confused, she definitely knew what was going on and listened to what was being said. Since she was so weak, she was talking even less.

I know it was probably irritating to my family, but I would leave the room every so often and go out in front of the hospital and smoke. Mind you, it was cold and nasty outside, but that didn't matter. I would smoke a couple of cigarettes, watch the people coming and going, pray, and try to gather strength to go back upstairs. The only thing I knew was that my insides were churning. I was so frightened. Momma was worse today. And they had hurt her! I couldn't stand this! My insides were on fire! I had to keep myself composed, yet I felt I would explode!

twenty-three

By that evening, I was beyond functioning. Or so I thought. Again I was in bed, but not asleep, when the phone rang. It was 9:30. It was Momma's doctor. She began by saying that the results of the biopsy were not good. It was malignant. I began to sob as she continued to speak. She said Momma had cancer of the gallbladder, which had spread to the liver. Although there were no signs of malignancy in surrounding tissues when she was operated on in October, the cancer was evidently dormant. Since the October surgery, a six centimeter malignant mass had grown across Momma's liver, sapping her strength. It was the reason for her poor appetite, weight loss, and deteriorating health. I have no idea what my end of the conversation was, because by this time I was just crying and saying "No!" The doctor said nothing could be done. They would keep Momma comfortable and medicate her if she had any pain. There would be no further tests. In plain and simple language, Momma was going to die. There are no words to explain the devastation of this phone call. None.

The doctor went on to say that she would tell Momma at seven the next morning. She wanted to know if I wanted to be there, or if other members of the family did. I didn't know how to respond. I couldn't absorb what she was asking me.

She said some people would rather the doctor tell the patient and others wished to be present. I remember crying, saying, "I can't. I just can't." She was kind and said she understood. I really don't know how we left things. I just remember I was actually in pain, my entire body, from head to toe. This lady had just told me that my Momma was dying, and there wasn't anything that could be done.

Gene and Kim were downstairs watching television. I hung up the phone and began screaming uncontrollably. They both ran upstairs. I was screaming and sobbing so hard that they couldn't understand me. Kim started crying, and Gene was holding me, trying to get me to pull myself together. They both had this shocked look on their faces. Kim was crying and kept saying, "What's wrong? What's wrong?" Finally, I got it out. I told them about the doctor's call. Kim started screaming and running through the house yelling, "No! They're wrong! No!" This was her beloved grandmother, and she was running from room to room screaming at the top of her lungs, "No!" She completely lost it, which pulled me back somewhat. I realized how she had been in complete denial. Gene was in a daze. I told him to go get Kim. By the time he brought her back in the room, I had pulled myself together enough to tell them that I had to go over and tell Marsha and Daddy.

Kim was still crying and saying No. Gene and I were going to go over to the house. I told Kim she couldn't go unless she got herself together. I told her I had no idea how I was going to tell Marsha and Daddy, but I couldn't have her there out of control. What I didn't say was that she was so upset I didn't want to leave her at home alone. We sat on the side of the bed and talked about how we all had to pull together. That's what Momma would expect. Gene talked to her some also, while I was getting dressed.

Since we had been at the hospital most of the day, all three of us ringing the doorbell at 10:15 certainly didn't mean good news. All Marsha and the kids had to do was look at our faces. Kim went back in the bedroom with her cousins. I got Daddy to sit down in the chair in the living room by the piano. Daddy started trying to make small talk with Gene. I interrupted and said I had to talk to him. Marsha was already in shock before I had said a word.

The Lord is always with us, because he allowed me to find the composure and strength to tell Daddy and Marsha what the doctor had said. I told them that the doctor would tell Momma in the morning and that she wanted to know if any of the family wanted to be there. Marsha quickly said she "couldn't, just couldn't." I said I understood and that I would go to be with Momma. Marsha and I were both crying, Gene was pacing the floor, and Daddy was just sitting there, I guess in shock. Then he started talking about bringing Momma home and how we would take care of her. It was obvious he was trying to put on a strong front for us, but his face was twitching. I knew I had just delivered a horrible blow to him. All along, he had been certain that Momma was going to get better. After all, she was the one who took care of him. He was the one who always talked about dying.

How was I going to stand there while the doctor told Momma she was dying? I felt like I couldn't take any more. But I couldn't let Momma be alone.

While we were in the living room, Kim told Kristy, Kevin, and Michael. She said that Kristy and Kevin were visibly upset and both began crying. Michael, unfortunately, internalized everything, so he just sat there, saying nothing. This didn't mean he wasn't in distress, because his grandmother meant the world to him, too. After all, he depended on her to make his applesauce every summer and freeze it for him. No one could duplicate Momma's secret applesauce recipe, and we all called it Michael's applesauce.

When we left, I assured Daddy and Marsha that I would be at the hospital in the morning before the doctor talked to Momma. Marsha said they would be over later in the morning. I promised to call.

This had to be the night from hell. I had been going from day to day thinking that things couldn't get any worse. As things stood, each day *was* getting worse, and I didn't know how I could stand it. The word *nightmare* doesn't even touch how bad things were for all of us. I could no longer convince myself that things were going to get better. This realization itself was devastating for me. I had, in my mind, been using this mental crutch to go from situation to situation. The crutch was gone. Momma was dying. The crutch was gone, and I had to face the reality of losing my Momma. How could I make it through this? How?

Friday, December 30, Gene and I were at the hospital at 6:45. We rushed to the nurses' station to see if Momma's doctor had come yet for her rounds. She had not yet arrived. Gene and I stood in the hallway outside Momma's room. There was no way I could go in, because she would want to know why we were there so early. Gene and I didn't talk. I kept watching the clock on the wall and praying for strength. I felt like my knees were going to buckle under me. I was about to pass out.

It was 7:18 when the doctor came around the corner. She

looked at us both and smiled pleasantly. I said something like, "I had to come," which I guess she acknowledged. It was still dark outside, so the room was dimly lit by a wall light. She turned on the light over Momma's bed, and Momma looked up at us. The doctor was on the right side of the bed, I was on the left, and Gene at the foot. I took Momma's hand. She said nothing. There wasn't even a trace of a smile. She didn't ask why we were there. The doctor said, "Harriet, I got your test results from the liver biopsy, and the mass was malignant." Momma didn't so much as blink. The doctor went on to tell her that the cancer was so advanced that they weren't going to put her through any aggressive treatment. That meant she didn't have to go through radiation treatment or chemotherapy. There would be no further poking and probing, and the goal was to keep her comfortable.

When she paused, Momma looked right at her and said, "How long do I have?" I then remembered that I had asked her that the night before. She really didn't answer me. She skillfully avoided a direct answer again, but as Gene remembers, she gave an outside time-frame of six months. I wasn't crying. I was frozen. For once, I had no words. None. She asked Momma if she understood, and Momma simply moved her head to acknowledge that she did.

When the doctor walked out of the room, I followed her, Gene staying in the room with Momma. I started crying as the doctor began to talk to me. She told me that the Hospice staff would be contacting me that day. They would move Momma to the Hospice Floor of the hospital where she would get more intensive care. She went on to say that she had never told a patient that they had cancer and seen a non-reaction like Momma's. She said she thought Momma knew this before they did. Her one concern was that Momma was so stoic. If she turned her face to the wall, it would take her out of here quickly.

That morning I called the church, only to find that the minister was in California with his family for the holiday. I told the secretary that Momma was seriously ill and that I wanted the church's prayers. Later that morning, the church Lay Leader, Brother Mills, and the Associate Pastor, Rev. Ms. Ruffin, came and gave Momma communion. Rev. Ruffin, in her late eighties herself, said she had been ill for several weeks, but when she heard about Sister Walker, the Lord gave her the strength to come to the hospital. This was very moving, because it was the dead of winter.

In the late afternoon, Momma still was mostly sleeping. Her sister Willa, Daddy, Kim, Marsha, Kristy, Heather, Aunt Willa's granddaughter, and myself were all there. Momma was aware of those present but was either too weak or just didn't want to talk. The lady came in from Hospice and asked Daddy, Marsha, and I to go down to the lounge with her. She explained their program. Momma would be assigned an RN and a health aide. They both would visit with her while she was in the hospital. This was in addition to the regular floor nursing staff. It seemed this lady and the young man social worker just went on and on. I probably didn't hear much, but I did manage to ask some questions. Momma needed 24-hour care. Neither Marsha nor I directly asked the question of how long did she have. But we both, in different ways and at different times, tried to get someone to vocalize this. This never happened, and I have no idea why it was even important.

Thinking back, it was so poignant that at the times when I could not function, Marsha picked up the ball and ran the other way around. Marsha couldn't go to the hospital that morning, and that afternoon I couldn't deal with these Hospice people. I was in and out of a fog. She took over and answered their questions and emphasized what type of care we wanted for our mother. Daddy was able to ask questions and was holding himself together much better than I was.

Then I snapped to when the lady asked if Momma had a Living Will. All three of us said no at about the same time. Daddy started telling how he didn't believe in that and that "No one was going to pull the plug on him or Momma." The lady quickly picked up that this was a sensitive issue and said we could decide that later. She needed to go back to Momma's room to get her to sign some papers.

When we got back to the room, I went in with the lady. Kim, Kristy, and Heather got up and flew out to the hallway with Marsha and Daddy. Marsha had said to me that she couldn't go in. Aunt Willa was seated in the chair. I went around to the left side of the bed and got up on the bed with Momma. This lady started telling Momma that they were moving her to Hospice and she needed her to sign some forms. She pulled out a clipboard and pushed it toward Momma. Momma just stared at her. I began to tell Momma that it was alright, that they were going to provide her with 24-hour care and make sure she was comfortable. I helped Momma sign her name on the various forms.

I was in tremendous pain, but somehow I didn't lose it in front of Momma. Then the lady said, "We'll talk later about a Living Will. If your heart should stop and there's no DNR (Do Not Resuscitate) order, then the staff must try to revive you. This is often brutal, and they sometimes fracture ribs." Momma just stared at her.

By this time, I was ready to jump across the bed and floor this lady. Maybe she was just doing her job, but I wasn't the least bit pleased about how she came across so direct and blunt. I ushered her to the door. Don't ask me what I said to Momma, because I don't remember. Aunt Willa just says she doesn't know how I did it. Neither do I.

That Friday night, I stayed until they moved Momma to the seventh floor. When the attendants left the room, Momma looked up at me and said, "Is this my final resting place?"

I quickly said, "No, Momma. You're in Deaconess Hospital, on the seventh floor."

"Where is that?" she asked.

"You know," I said, "right across from the University."

All she said was "Oh." To this day, I don't know how I didn't have a stroke right on the spot. The Lord is the only one who helped me respond to Momma, calmly and with compassion.

I now think I understand her confusion. For years, there was only one hospice in the city, on Reading Road. This was something new, each hospital having its own hospice unit. I think Momma thought she had been moved to the original hospice, when she had only been transferred from the fifth floor to the seventh.

As I drove home that evening, my mind jumped to Daddy. I had been so absorbed in my upset and pain that I hadn't even considered what Daddy must be going through. I was thankful Marsha was with him and was suddenly aware that I needed to spend some time with him also. How, I didn't know, but I needed to give him more attention and support. And then there was Aunt Minnie. She had to be told about Momma. How was she going to handle this? I had no idea. I just knew it would be hard on her.

That evening, I was beyond thinking. I was in another world and had no idea how I was going to do whatever had to be done. I looked at all the Christmas lights on my way home and began to cry. This holiday had been a nightmare. And tomorrow, Saturday, was New Year's Eve. A new year was dawning—1995.

twenty-four

During that day, at least three nurses talked to Marsha and myself about signing a Do Not Resuscitate order. Each explained that resuscitation was a brutal procedure and that, if the patient was revived, they were then put on a ventilator. Neither of us could deal with this, so we put them off. Quite frankly, I didn't understand why it was so urgent for us to decide.

That evening, I got a call from our cousin Lorna, from the hospital. She was there and had talked to one of the doctors. In fact, I think it was Momma's surgeon, because she had been with me when Momma had surgery. Lorna was crying and said the doctor told her to tell us to sign the DNR order right away. She said we should call the nurses' station.

I called and got both Marsha and Daddy on the phone. I explained, the best I could to Daddy, what everyone had said about the procedure. I told him that the doctor had promised to keep Momma comfortable and that was what Momma wanted. Daddy got very emotional and wanted them to do everything possible to keep Momma alive. He just didn't believe in pulling the plug. It should be God's work, he'd say. Marsha chimed in and we both agreed with him. She tried to explain that this was not "pulling the plug," only that they wouldn't try to revive her if her heart stopped.

After what seemed an eternity, with all of us crying, Daddy said it was up to Marsha and me. He didn't want to have anything to do with it. I called the nurses' station and told them they could put the DNR on Momma's chart. We were again torn up making this decision.

The next couple of days were tough on everyone. Kim changed her flight to the following Wednesday. Marsha's children would drive back to Maryland on Monday, and Marsha would stay at least for that next week. Marsha had to deal with the phone ringing at the house and explaining to people about Momma's condition. This had to be very tough on her, as well as on Daddy.

Sunday evening, the four grandchildren came to the hospital. Although Kevin had driven, something told me I should be there, too. When I arrived, Kim, Kristy, Kevin, and Michael were all in the room watching the football game. Momma was in and out of sleep. I sat on the side of the bed and talked with the kids. Suddenly I realized that they didn't want to leave and say good-bye.

Visiting hours were over, and it was about 8:30. I told them we had to go so Momma could get some rest. No one said a word as they put on their coats. Then Kevin said, "Bye, Momma. I love you." Momma looked up at them and said softly, "I love you all, too, and you've made me so happy." I think they each kissed her, but I'm not sure because this sent me into orbit.

When we got out into the hallway, we stood in front of the elevator sobbing. This was so painful for us, because the unspoken was that they wouldn't see their grandmother alive again. I put my arms around Kim, Kristy, and Kevin, all of us really crying. Suddenly I looked up and Michael was standing over looking out the window. I called him and went over and held him, too. Michael was grieving as much, if not more, than we were, and he needed our support.

Monday night I called the nurses' station around eleven o'clock to check on Momma. The lady who answered said her nurse was in Momma's room and that she would transfer me. Now I wasn't calling to talk to Momma, I just wanted to know how she was doing. The nurse answered Momma's phone and said, "Oh, she's fine. Here, you can talk to her." Momma was too weak to talk. I told her I was just checking on her, and Momma said, "Please come and get me! I want to go home! I'm afraid here. Please come and get me!" I could hardly hold myself together. It was like being pierced by a sword. I said I couldn't come and get her tonight, but if she wanted to come home, then we would bring her home. She sounded so frightened and pitiful. My heart was aching, because she was begging me. I told her I'd be there in the morning and find out what we had to do to bring her home.

When I hung up the phone, I cried and cried. Both Gene and Kim tried to talk to me, but I felt like I had really let Momma down. Kim was crying, too, both of us ending up with headaches.

Marsha and I met with the Hospice nurse the next morning and made arrangements for Momma to come home on Thursday. They had a hospital bed delivered, all the necessary medical supplies, and arranged for 24-hour nursing and ambulance transportation home. I was petrified, because it was the coldest week of the winter. I didn't know if Momma could stand to be moved—she was so fragile. But I wanted her home, and Daddy was thrilled.

That Tuesday night, Gene, Kim, and I went to the hospital. Kim was going back to St. Louis the next morning. Momma wasn't talking much at all. It had become too much of an effort. When we left, Kim kissed her and told her goodbye. Momma didn't say anything.

When we got out in the hallway, Kim completely lost it. She sobbed that Momma didn't love her. She had told the other grandchildren that she loved them and that they made her happy, and she didn't say anything to her. Kim was almost hysterical. Both Gene and I held her. "Kim, Momma isn't talking now. You know she loves you," I said. "Don't be silly. I was here the other night when she said that, and she was talking to all four of you." Gene said he understood how upset she was about her grandmother but that there was no way Kim could believe Momma didn't love her. It took us some time to pull ourselves together from yet another painful moment.

Thursday, I arrived early and sat with Momma, telling her over and over that she was going home around noon. I kept explaining that she would ride in the ambulance and that I would drive my car. I was worried about her being frightened, like when they moved her between floors. She wasn't saying much, so I wasn't sure she fully understood, but she wasn't agitated. I told her about the hospital bed and that Marsha and Daddy were at home getting her room ready.

Just as the ambulance drivers arrived with the stretcher, I heard a voice in the hallway say, "I'm looking for Harriet Walker." I went over to the door, and here came Momma's good friend, Mrs. Keys, down the hallway. It was zero outside, and she was bundled up like she had come from the North Pole. Her face lit up when she saw me. She had taken a cab from her home, about twenty miles. She wanted to see

Harriet and bring her some talcum powder. The men were getting ready to lift Momma onto the stretcher when Mrs. Keys told them to stop. "Just wait until we have a prayer." We all bowed our heads as she prayed. She kissed Momma, but Momma wasn't able to speak. I believe Momma knew that her friend was there. I felt terrible that I couldn't take her home, but I had to go with Momma. Mrs. Keys said she understood and would get a taxi back home. She was happy that she didn't miss seeing Harriet.

There's no question in my mind that Momma knew she was home. Gene and his friend Jack had taken down the double bed in the guest room, making room for the hospital bed. Marsha had made the bed and put on the Dogwood quilt made by Momma. On the dresser, she had put pictures of all the family, and the room was bright and cheerful. My worries about the cold had been unnecessary. Momma made the trip fine. She was happy to be home. We all believed she knew she was home.

twenty-five

By Friday, Marsha had definitely decided to stay for another week. She just couldn't leave Momma. Even with the nursing help, it took both of us. But more importantly, Momma needed us both, and we needed to be with her as long as we could. Neither one of us had any idea how long Momma could hold on. She was on oxygen, but wasn't taking any medication. It was a blessing that she wasn't in physical pain. As a matter of fact, other than immediately after the surgery, Momma had not suffered with pain. She was just weak.

While in the hospital, Momma had received fluids and supplements through the IV. That was how she was getting nourishment. When I asked the nurse about her not getting the IV at home, she just said, "Get her to take as much fluids as you can by mouth." I turned and looked straight in her face and said, "She'll starve to death." The nurse had this blank look on her face and repeated, "Just give her as much fluids as you can by mouth." I couldn't believe it. How were we going to get Momma to eat?

The first couple of days, Momma ate some applesauce, and we spoon-fed her soft foods. We spoon-fed her Ensure and ice chips, too, because she was too weak to sip from a straw. She had a catheter, so the nurses monitored her urine output. She had to be turned every couple of hours, too, and rubbed down with lotion.

A couple of days before Momma left the hospital, the nurses had taken out her teeth, saying her mouth would be more comfortable and easier to rinse out. Momma didn't object, and was still without her teeth. This says a lot about her condition, because Momma would never let anyone see her without her teeth.

I would arrive each morning and stay until the nurse came at 11:00 p.m. Marsha and I managed to stay busy doing things around the house, talking to Daddy, answering the phone, and checking on Momma. Saturday, our cousin Roger and his wife Evelyn came from Indianapolis to see Momma. Not only was Roger Momma's nephew, but he had been one of her second-grade students. Now he had retired from the military and was getting ready to retire from Ameritech. As we all stood around Momma's bed, I said, "Momma, one of your second graders is here to see you." Momma could no longer smile or acknowledge our presence, but there was something about her eyes that made us feel she knew what was being said.

One evening, the nurse didn't show, and Marsha and I tried to change Momma's gown and bathe her. As it turned out, we were doing it all wrong. We had raised the bed, instead of lowering it. I told Momma I was going to roll her over toward Marsha, and Momma tried to put her hand on that side of the bar. While we were turning and juggling her around, I said, "Momma, I bet you're sorry one of your daughters isn't a nurse. We sure don't know what we're doing. I hope we aren't being too rough." Of course, there was no response, but when we said, roll toward Carol or Marsha, she attempted to do so. To us, that meant she was aware.

On Sunday, January 8, several people came by the house after church. Cousin Iris, her daughter Kim, and Kim's new baby, Iris Patricia, came by to see Momma. Iris Patricia was the newest member of our family. Momma had her picture on the mantel, but had never seen her. Momma just stared at them, but again we felt she was aware of who was there and what was being said to her.

Daddy spent most of his time sitting in a chair next to the bed, holding Momma's hand. Often, we'd go in and he would have his head on the bed rail, asleep. The nurse would sit over in the rocking chair, reading a book. The only noise was the oxygen machine going on and off. Momma mostly slept, except for when we were trying to punch liquids or ice chips in her mouth.

When I arrived on Monday, there was a new nurse. From the instant I met her, I didn't like her. She talked too much. She came across to me as a know-it-all. The other nurses had just gone about their job and pretty much faded into the background. There seemed to be no intrusion in the household. But this lady was different. She wanted to come in the kitchen and talk rather than stay with Momma. It was as if she had a need to impress us with her medical knowledge. In short, I was turned off by her.

I decided on this day I would clean out Momma's china cabinet and clean the silver. I had told her before she got sick that I would help her do this, as well as clean the chandelier. I had cleaned the chandelier on Saturday, so today I would tackle the china cabinet. This was a major job, but each of us had been doing things to keep ourselves occupied. Marsha was doing something in the basement.

By the time the hospice nurse, Jan, came in the afternoon, I had the kitchen counters and table full of sparkling crystal and silver. Jan had been there on Thursday, when Momma came home, to see that everything was set up and ready for her. She was there Saturday and phoned daily, talking to the nurse or aide on duty. She had spent time with Marsha, Daddy, and myself, letting us unload. I must say, one truly felt her support. She told us that if something happened, the DNR form must be taped to the back of Momma's bed. If someone called 911, they would have to resuscitate if this order was not present. The orders were that she was to be called or paged by the nurse if there were any changes, and she would come immediately, regardless of the hour.

Out of only these few encounters, I had come to have great respect for this stranger. She had moved into our lives at this time of need. She showed compassion and understanding and was calm, gentle, and nonjudgmental. She didn't know any of us, yet said she could feel warmth and family just walking into our home. Jan really brought me comfort, and I was so very much in need.

After Jan checked on Momma, she talked with this nurse I didn't care much about. Then she came into the kitchen, and we sat down amidst the dishes. She called the registry and found that there was a problem with the 3:00 to 11:00 nurse. The weather was terrible. They were having trouble getting assignments filled. She went back into Momma's room and talked with the nurse. This lady agreed to take a double shift, so that was better than Marsha and I struggling on our own. Even though I didn't care much for this lady, we needed help. The 11:00 p.m. nurse was scheduled, and so we had coverage lined up for the next couple of days. Jan told us that

Momma's vital signs were better than they had been on Saturday.

I went in and told Momma that she would be pleased to know that not only had I cleaned the chandelier on Saturday, but that I had just finished cleaning everything in the china cabinet. I said something like, "I'm sure that will make you feel better." Now, I was talking to my mother. So this nurse gets up from the chair and says in front of Momma, "You know she's dying. See her legs. She's beginning to mottle."

I was in a total rage. As I walked out of the room, I said, "Please come here." We went into the dining room, and I turned and told her, "Don't you talk that way in front of my mother!"

She apologized, but went on about when your legs start turning purple in splotches. It's called mottling, though at the time I thought she was saying "modeling." This mottling was a sign that dying was imminent. There was no way I wanted to hear this and not from this lady. As a result of this irritation, I pretty much avoided this nurse all day, which meant I didn't spend much time with Momma that day either.

That evening, I decided to go home early. It was around seven, and of course I was tired. When I got ready to leave, the nurse had the door to the bedroom closed. She was evidently changing Momma's gown, so I didn't bother to go in and tell Momma I'd see her in the morning. I just went on home.

At exactly 10:50, Marsha called. "I don't want to frighten you," she said, "but Momma's breathing has gotten shallow." That's all she said. "We're on our way," I said. I screamed for Gene, put on my sweat suit, and we were out the door in minutes. I was crying and Gene was trying to reassure me that things would be alright. It seemed we were creeping along, but Gene was actually speeding.

It was 11:20 when we pulled up in front of the house. I was out of the car, running across the lawn, before Gene got the motor turned off. As I got to the door, Marsha opened it, Daddy standing behind her. They were both crying. "Momma's gone!" Marsha said. I screamed "No!" and ran toward the bedroom. The door was half-closed. I pushed it open, sobbing and calling Momma. The nurse moved aside. "Momma!" I cried. I grabbed her hand, touched her face, and she was gone. Her body was still warm, as if full of life, but her breath was silent. Her face was as smooth as a newborn baby's. There were no wrinkles, no twisted left side from Bell's Palsy. She was absolutely beautiful. Daddy was behind me saying, how beautiful Momma looked, how she looked the way she had when he met her. In my unbearable grief, I saw a peace on her face that I shall never forget. It was truly well with her soul. There is a God and there is a Heaven and Momma arrived there at 11:08 p.m. on January 9, 1995.

~

Momma couldn't be declared dead until the hospice nurse, Jan, arrived. So as a matter of record, her death was noted as 12:05 a.m., January 10.

Jan and the duty nurse took charge. The funeral director was called, and Momma's death was reported to the coroner's office. Daddy, Gene, and I stayed in the kitchen, and Marsha went down in the basement. Gene went to check on her, but she just needed some time alone. When Marsha came back upstairs, we made a few calls. I remember calling Aunt Willa, Momma's friend Pinkie, and I don't know who else. We were all like zombies.

Gene was with the men who had come from the funeral home to take Momma's body. Daddy, Marsha, and I stayed in the kitchen until they had removed her remains. Jan came in and talked with us, then gave the funeral home men all the information they needed.

After they left, Jan came to me and said she was required by law to count Momma's medication in front of me and then to flush it down the toilet. I said Momma hadn't taken any of the pain medication. She told me that the afternoon nurse had given Momma pain medication twice, according to Momma's chart. The chart said Momma was agitated and possibly in pain. Neither Marsha nor I had been aware of that or that Momma had been given any pain medication. But, sure enough, when the pills were counted, two were missing.

It was about 3:30 in the morning before Gene and I went home. My mother of 51 years was gone. My pain was unreal. My family's pain was unreal. This just couldn't be happening. Not to Momma!

Thinking back on that night, I now believe there were:

Four Angels at Momma's Bed,
Two at the Foot,
Two at the Head,
And the Four Carried Her,
When She Was Dead....

twenty-six

By the time I arrived the next morning, Marsha had already made most of the necessary calls. The funeral director had called and would come out to the house the next day, Wednesday morning. We had agreed to have the funeral on Saturday morning. It was on this day, Tuesday, that Daddy started saying every few minutes, "I just don't know." He still says that, to this day.

We had called Kim the night before, and she had gone to pieces. She couldn't wait to come with Toren for the funeral, so she got the first flight Tuesday morning and Gene went to pick her up at the airport. Kristy, Kevin, Michael, and Ronnie would fly in on Friday. While Marsha and I sat at the kitchen table, we thought of our cousin Raymond, in California. I picked up the phone and called. I was shocked to find out that he already had a plane ticket for Wednesday. He was coming to see Momma. Now it would be for her funeral.

By mid-morning the phone was jumping off the hook. Word travels fast, and people were calling to express their sympathy. I wasn't aware of it at the time, but this was another time that Marsha came through when I fell short. It seemed that everyone who called had to tell you how much Harriet meant to them. They would go into great detail, telling what a wonderful person Momma was and how close they were to her. During these conversations, all I could say was, "Yes, I know." Now, don't misunderstand me—these calls were wonderful tributes to our mother, and there were many of them. But I was grieving so much that I couldn't handle it. So Marsha stepped in and took over the calls, and I was so grateful.

When the minister came to the house, Dr. Clark, I withdrew. He had only been at our church since August. Although he married Kim and Toren in September, he had little knowledge of our family, nor of its contributions to the church. As we gathered in the living room, he asked that we tell him something about Momma. He had visited her once in the hospital in October. Inside, I felt so hurt that he had to ask us to tell him about our mother. Again Marsha rose to the occasion and was eloquent in describing every detail of Momma's church, community, and family work. I sat there saying nothing. I couldn't contribute. I just couldn't tell him about Momma because I felt he hadn't taken any time to get to know her. I remember at one point I left. I went to the kitchen and wrote out some bills, leaving Marsha and Daddy in the living room. Before the minister left, he came into the kitchen to say something to me. I guess my leaving the room was rude, but I had to remove myself and focus on something else.

That afternoon, Kim and Marsha went to the nursing home to tell Aunt Minnie. I couldn't do that either. Aunt Minnie was really shaken by the news. Marsha said that after she had collected herself, she had to go tell everyone about her sister-in-law. Oh, she didn't forget to send me her laundry either.

One thing I did manage was to write Momma's obituary.

Marsha put the program together, selecting the scriptures and songs. We decided that Momma would wear her beautiful pink sequin dress, the one she bought for their fiftieth wedding anniversary. She really loved that dress, and we would have a casket spray of pink roses. This was all very fitting, since Momma was a member of Alpha Kappa Alpha Sorority and their colors are pink and green. In fact, not only was she a member, she was a "Golden Soror," which meant she had been a member for over fifty years. Of this she was quite proud.

Family and friends were very supportive. Gene's brothers and sister sent over a ham and a turkey. Food was coming in from neighbors, as well as from others. Flowers were arriving, even though we had requested that contributions be made to the Mt. Zion Methodist Church Scholarship Fund. People were kind and generous.

One call I did take was from the American Cancer Society. Momma was the Area Chairperson for the Region. She had done this for years, collecting the money for her area, a Brinks truck coming then to the house to pick up the money. The lady wanted to speak to Harriet about the April Campaign. I told her that Momma had just passed, and her voice trembled as she expressed her sympathy. The next day we received a beautiful card from the American Cancer Society and a single red rose in a bud vase from Hospice Homecare.

On Friday afternoon, Momma's doctor called. I answered the phone. She expressed her sympathy and asked how we were doing. She went on to say that all of the doctors thought Momma had at least six weeks. She was surprised, and yet she felt that this was Momma's will. She again discussed Momma's non-reaction when she told her and said she believed that people could will themselves to die. She felt Momma would not want anyone having to care for her. I believe there is some truth to this. Momma was the care provider, that was her role. She always said she hoped she never had to go a nursing home. Her wish was granted. Momma left this earth of ours with great dignity and peace.

Saturday, January 14, was very cold, but was bright and clear. Visitation was from 9:30 until 10:30 and was followed by The Optimist Welfare Club service and the Alpha Kappa Alpha Sorority service at 10:45. The funeral service was to start at 11:00.

It was unbelievable the number of people who came out to express their sympathy and pay respects to our mother. In the front row were: Daddy, Marsha, Gene, myself, Kim, and Toren. In the second row were: Kristy, Kevin, Michael, Ronnie, and Marsha's ex-husband, Steve. Behind us was the rest of the family, which took up almost three-fourths of the left side of the church.

Both the Optimist and AKA services were wonderful tributes to our mother. Each read a resolution that truly captured the essence of her dedication and service to others. These resolutions along with the one read from Mt. Zion United Methodist Church, by Mrs. Robinson, left nothing else to be said. They were all magnificent tributes.

During the Alpha Kappa Alpha Service, it is customary for a member to place an ivy sprig in the casket. Since so many of my family members are in the sorority, they had each of them put a sprig of ivy in the casket, including Marsha. On that day, at that moment, I was truly sorry I hadn't joined the sorority. In their white outfits, her sorors turned out in number to pay tribute to Momma. They formed a horseshoe in front of the casket and around the sides of the church.

The Celebration of Momma's Life started at 11:00. Marsha had selected the scriptures and the congregational hymn, "Great is Thy Faithfulness." Shannon, a member of Mt. Zion, had grown up in our church and, like so many others, had been in Momma's Sunday School class. Shannon attended Boston Conservatory and was studying Opera. She came home for the service. We were so honored to have her sing "It is Well With My Soul" and "The Lord's Prayer." As her voice rang out like that of an angel from above, I looked to the picture of Jesus above the pulpit. My entire body shook uncontrollably. A flood of tears ran down my face. I wasn't aware of anything but my heart-wrenching grief over losing Momma. I knew without a doubt that she was in God's Kingdom, but I selfishly didn't want her to leave us.

As we filed out of the church, I then realized how many people were there. The church was filled. We had paid for a police escort for the procession to the cemetery. It wasn't that far, but there were so many cars that our driver said we should have had two escorts. When I looked back, the procession stretched as far as I could see.

After the grave site service, the family went back to the church for a luncheon. This had been arranged by Momma's Circle Six. Other members of the church helped out, too, serving a full turkey dinner. I couldn't eat, so I just went around thanking people for their kindness. I was numb. I was in a complete blur and just wanted to wake up and find that this had been a nightmare.

That evening, following the funeral, we were all at the house, five-year-old Ronnie running around trying to cheer everyone up. He couldn't grasp what was going on. He was running through the dining room when I grabbed him, telling him to slow down. Sticking from his pocket was a crumpled pink rose with only one leaf left. He told me this was his rose from Momma's "funeral," for him to keep. I just stood there. None of us had seen him get the rose from the floral spray on the coffin. He had put it in his suit coat pocket. I was so touched.

When they got ready to go to the airport the next evening, Ronnie came flying back in the house, saying he had forgotten his rose. He found it, and off he went with the pink rose smashed in the palm of his little hand.

Marsha's ex-husband, Steve, had driven down from Maryland. Although they had been divorced for a number of years, he had great respect for Daddy and Momma. Despite their divorce and his re-marriage, Momma and Daddy always treated him the same, and whenever they were in Washington, he made a point to go see them.

Raymond, our cousin, who had come from California that Wednesday, stayed for several days. He took time to come by the house and sit and talk with Daddy a couple of times. Jackie had flown into Columbus from California and came down with her parents for the funeral. This meant so much to me because she had just been here the day after Christmas, with her husband Ron. So you can see that Momma touched a lot of people and left each with special memories.

The grandchildren all left on Sunday, and Marsha stayed until Monday night. We tried to convince Daddy that he should come and stay with me. This turned out to be wasted energy. His mind was set that he would remain in his house, and that was that.

twenty-seven

Aunt Minnie had made it very clear that she wouldn't go to the funeral. I respected her wishes and didn't insist. She'd start crying about Harriet, and then I'd start, so this was just causing more pain. I promised to bring her a couple of programs.

When I went by the nursing home on Tuesday, not only did she have her laundry for me, but she also had about twenty copies of Momma's obituary from the two local papers. She had them neatly cut out and in a plastic bag. She was so proud, because both articles were quite lengthy and both had Momma's picture. Evidently, she had gone from room to room getting people's newspapers. Bless her heart, she was so proud of these articles and wanted me to have copies to send to people. I realized that this was one way she could handle her grief and feel like she was helping. I thanked her over and over, and she was so pleased.

Back in October, when Momma came home from the hospital, we had arranged for her and Daddy to get their evening meal from Meals on Wheels. I arranged to have this continue for Daddy. That meant that in the morning he would have to get his cereal, banana, and milk. While Momma had been sick, I bought all kinds of food in hopes that she might eat something. So there was a lot of stuff there in the house. Getting him to fix something for himself was the hard part. Even peanut butter and jelly. Yet, he said I wasn't supposed to worry about him!

The Lord works in mysterious ways. The week after Momma died, there was a card in the mail from one of the ladies who had taken care of her several times during her last days at home. Her name was Lu, and she had been just wonderful with Momma. She sent an inspirational card and message to our family, and I was so touched that I wrote her a thank-you note and sent her a check. In the note, I asked if she knew of anyone who could help with my Dad. She responded and said that she would be willing to help out.

Lu has been a true blessing to our family. She comes once a week and does the housekeeping and laundry. This relieves me of this work, and I'm most appreciative and grateful. Daddy has really bonded with her, too, and looks forward to her coming. He gets her to scramble eggs for him—like I couldn't do this for him!

Fortunately, both Momma and Daddy had wills. They both had always been very secretive about their business. Even though I had been doing their taxes since 1989 and knew where they had accounts, that was about it. Daddy, however, knew every detail. This had been his job, handling the finances and paying the bills, until he lost his vision. Momma was never interested in this and could never satisfy him once she had to take over. Daddy knew when everything was due and wanted bills paid the day they arrived. Momma would put them on the kitchen window sill and say, "I'll pay it later. They don't need the money before the due date." This was always so funny.

Now I had to sort through all this stuff, going back and

forth between the Hamilton County Recorder's Office and Probate Court to get things settled. It was time-consuming, but we saved a lot of money in attorney fees. Daddy was very appreciative and looked forward to a report each day. We had several arguments because I insisted on combining some accounts. They had money spread out all over, and I needed to make their affairs less cumbersome. This was difficult to get across, but Marsha supported me and finally Daddy agreed.

So after weeks of working, visiting the nursing home, checking on Daddy, and pulling together their business, I finally had things in order. That is, everything except the property. I got Daddy to agree to hire an attorney to take Momma's name off the deed and to revise his will. The attorney also drew up papers to give Marsha and myself joint power of attorney. Boy, was this traumatic. Daddy refused to sign a Living Will prepared by the attorney. This whole thing became very traumatic for him. I think he equated it with losing control. No matter what Marsha and I said, it didn't make him feel any better. But he had to face the reality that he needed help with his affairs. The attorney did a good job of trying to help him understand and explaining options, and both Marsha and I got power of attorney.

Another big issue was Momma's china, crystal, and silver. Marsha, Daddy, and I agreed that we needed to go through the china cabinet and pack up the things that were valuable. In fact, Daddy even said, "Someone could come in and walk out with the refrigerator, and I wouldn't see them." Of course, this was an exaggeration, but it led us to believe that he was in agreement about putting away anything valuable.

One evening Gene and I came in with boxes to pack some

of these things. We had already said we would leave things in the cabinet, not strip it bare. Gene is an antiques buff and particularly knows silver and glassware. I was in the kitchen talking to Daddy when Gene started wrapping pieces in newspaper, putting them in boxes. Then I got up to help. Daddy came into the dining room, sat down and started crying and fussing that we were tearing his house apart. Gene and I both stopped dead in our tracks, shocked. I again explained what we were doing, and it was as if he had never heard this before. He was so upset that we just taped up the boxes and put them in the back bedroom, stacking them in the corner.

We assured Daddy that we weren't taking his things, just trying to protect them. Later, I found out that Daddy called Marsha and told her we were taking Momma's things. Marsha went through the whole thing about our agreeing to put some things away. He evidently didn't want to hear this from her either. It was just too painful for him to see anything moved, so we didn't move anything else. Not even Momma's 1992 Buick Century parked in the garage.

twenty-eight

Although I was functioning, doing all the things that needed to be done, I was hurting so bad. I missed Momma terribly and either said nothing to people or talked too much about her. I had my mother for 51 years, and now she was gone. Things would never be the same. I was mentally, physically, and emotionally exhausted.

On Thursday evening, Aunt Minnie called me at home. She wanted to know when I was coming over, because I hadn't been there that week. "I need some clean clothes," she said. I chided her, saying I knew she had enough clean clothes and that she just missed seeing me. I told her how busy I'd been and that her clean clothes were in the back of my car. I said I'd be over that weekend, "...just like the mailman, neither rain nor sleet nor snow...." We both laughed.

The next day, Friday, February 3, I stopped by to check on Daddy after work. I usually waited for him to eat before I would leave, but on this particular evening, we were to get a snowstorm. As I sat talking to Daddy, I looked out the kitchen window and realized it was starting to come down mighty fast. It was about 5:15.

Daddy insisted that I go home. Not only was it snowing, but the temperature was around zero. I bundled up and left within the next ten minutes. Well, it usually takes me thirty minutes to get home, but on this evening, it took two hours. The roads were terrible. The snow was turning to ice as soon as it hit the pavement, and it was snowing so hard I could barely see. I was a total wreck driving on the slick streets. I had two bad accidents on icy, snowy roads several years before, and ever since, I've been nervous about driving in that kind of weather. So I was extra tense.

When I walked through the door, I called Daddy. He couldn't believe it had taken me that long. I hadn't called him from my car because I didn't want him to worry. I told him I had a headache and was going to bed. My entire body was tense, and I couldn't shake it.

By 9:30, I wasn't asleep, but I was in the bed. It was still snowing outside and everything was sparkling white. This was now fine with me since I was home. The snowplow had been down our street maybe an hour before, but the street was already covered again. Looking out our bedroom window, you could see snow-laden trees across the hillside. It was beautiful! Thank God I had gotten home safely, I thought, as I looked at the peaceful, falling snow.

The phone rang at 9:45, and I answered. It was a nurse from the nursing home telling me that Aunt Minnie had suffered either a heart attack or a stroke. This had happened at 9:30, out in the lounge. The life squad was there working on her. They were unable to get a heartbeat. I was holding onto the phone, in total shock again. The nurse said she would call back as soon as the paramedics got her stabilized. I hung up the phone and yelled for Gene. He came flying upstairs, and I told him about the call as I got dressed. I kept saying over and over, "Aunt Minnie didn't have heart trouble." And she didn't. Her pressure had never been really high, just borderline high, and this was controlled by medication. She took that one pill, and that was all.

The phone rang again, and I grabbed it. The nurse said the paramedics still hadn't gotten a heartbeat. They were transporting her to the hospital that was, literally, around the corner from the nursing home. I asked what had happened. She said Aunt Minnie had been standing at the nursing station for about ten minutes, talking and laughing with them, then told them she was going to the lounge to watch the basketball game. She went out to the lounge, sat down on the couch, and slumped to the floor. Another patient called a nurse over. This had all happened at 9:30, and the life squad was there within ten minutes. By this time, I was all to pieces, but I thanked her and said we were on our way to the hospital.

Gene drove, and we had to go slow because of the bad roads. It seemed we were creeping. The only good thing was that the roads were clear of traffic. People were home. It took us about twenty minutes to get there. When we pulled into the emergency lot, the ambulance was still sitting at the entrance with the door open. We went flying in through the emergency exit. I was beginning to lose it big-time by then. I saw a man in green scrubs look at us and close his door. I approached the desk and said, "I'm Minerva Williams' niece. She was just brought in, and I want to see her." This young doctor walked over to the desk and said, "I just got her heart started. You'll have to wait in the waiting room until we know more." He turned and walked away. I just stood there with my mouth hanging open! This lady got up and showed us to the waiting room.

By this time, I couldn't stop crying. Gene once again was telling me not to jump to conclusions, that Aunt Minnie was really tough! All I could do was sit there and keep saying that her heart must have stopped for at least forty-five minutes. This was based on what the nurse at the nursing home had said and the doctor now saying they had just started her heart. Common sense says that is serious. Gene paced the floor as I kept saying, "I can't believe this! I can't believe this!" We were the only people in the emergency lounge.

After about twenty minutes, the emergency room doctor came into the lounge area. Both Gene and I were sitting, and he stood in front of us and told us about Aunt Minnie's condition. He said the paramedics had tried everything, but couldn't get her heart started. When they arrived at the hospital, he gave her something as a last ditch-effort, and her heart finally started. She was now on a ventilator. He was unsure whether she had a stroke that caused a heart attack or a heart attack that caused a stroke. Her condition was critical, and he didn't know if she would make it through the night.

No! No! No! I was crying as he talked, but I heard every word. There was an arrogance about this young doctor, that really turned me off. Later, I would find out that it wasn't just me; Gene felt the same. Someone needed to teach him how to communicate with compassion. In fact, it should be part of a physician's training, and if they can't cut it, then they shouldn't deal with people. This doctor said something like, "She's probably brain-dead, and if it was one of my family, I'd just let her go!" Then he said we could go back and see her. I jumped up, not knowing what to expect. We followed him into a room off the emergency entrance, the same room where I had seen the man close the door when we came in.

Words can't begin to describe what Gene and I saw that night! Aunt Minnie had tubes coming out of everywhere.

There were several IV's and all kinds of equipment, including the ventilator, which was breathing for her. The most gruesome thing was the way she looked. Her complexion, usually a warm, reddish brown, was now an ashy color. I couldn't recognize her as my Aunt Minnie. She was blown up like a balloon, and it looked like her skin was ready to pop. Her facial features were unrecognizable because her nose was spread across her face, her lips swollen up like balloons, and her forehead twice its normal size. Aunt Minnie looked like a monster from outer space. My beautiful, funny, warm, caring aunt. This was unreal, truly unreal.

I will never forget that moment, and I resent that no one prepared us for what we were about to see. We should have been told that, in trying to resuscitate someone, fluids and medications are given that distort their features and make them look like this. I was destroyed to think that this was my beautiful Aunt Minnie lying there looking like a creature from another planet.

As I cried and talked to her, I was angry, upset, scared, and a zillion other emotions ran through my entire body. I, once again, within a period of weeks, found myself helpless. This seemed too much to bear. They asked us to go back to the lounge and wait. They were running tests and would let us know if she stabilized. The doctor didn't expect her to make it through the night.

It was around midnight when we went back to the lounge. Outside, snow was still falling, and it was quiet as could be. No traffic, no noise, just softly falling snow covering everything. I went outside and stood, smoking, trying to collect myself so I could call Marsha and Daddy. As I looked around, the silence of the night was awesome. Everything looked like a winter wonderland. I couldn't believe I was there, at that place, at that time, with my Aunt Minnie lying inside critically ill. I tried to tell myself that she was a fighter and would pull through this, but I had seen her. I knew I had to call and wake up Daddy, because she might not last the night. How was I going to tell him? He still couldn't deal with losing Momma. This was his last sibling. How in the world could I tell him, especially when I couldn't even be there with him?

Gene agreed that I needed to call Daddy. This was one time he wasn't telling me not to jump to conclusions, because he, too, had seen Aunt Minnie. I went to the pay phone and called Daddy. The phone rang probably thirty times before he answered. This is another problem, that he can't hear. He was groggy but knew it was me and wanted to know what was wrong. I told him that Gene and I were at the hospital in Kenwood and that Aunt Minnie had been brought there from the nursing home. I told him that they didn't know yet whether or not it was a heart attack or a stroke but that it was serious. I told him I was really worried about her and that Gene and I thought he should know. Daddy was very calm and told me he was sure she would get good care. He sensed my upset, even though I wasn't crying. He told me not to stay there all night and to keep him posted. I never told him that they didn't think she would last the night, nor did I tell him I was going to call Marsha.

After hanging up from Daddy, I called Marsha in Maryland. When she answered the phone, she immediately wanted to know what was wrong. I could hear the panic in her voice. That's all it took for me to break down and start crying, telling her that it was Aunt Minnie. Sobbing, I told her all the details. She was shocked but made every effort to comfort me. She knew how close I was to Aunt Minnie and that this was too much for me after our loss of Momma. We talked until I was able to pull myself together. I was a complete wreck. It was another night from hell. I truly felt I couldn't take any more.

When I went back to the lounge, Gene was standing looking out the window. We were the only people in emergency that entire night. It was very quiet with only a few staff members passing by from time to time. Gene went and got me a cup of coffee. Again I pulled myself together. A lady from admissions came in and asked me to come to the desk to give them the necessary information about Aunt Minnie. Since I had been paying her nursing home and medical insurance for the last year, I was able to tell them what they needed. I was listed as the next of kin, even though legally it was Daddy. Since I had power of attorney for Daddy, I could act in his behalf.

A little after 2:00 a.m., the doctor with the not-so-wonderful bedside manner came and told us that Aunt Minnie was stable. They would be transferring her to Intensive Care. He suggested we go home. He indicated that the ventilator was keeping her alive. A house physician would be assigned to her, because the physician that she saw at the nursing home was not on staff at this hospital. He said we get could get more information in the morning. Believe it or not, I thanked him. It must have been my upbringing coming through, because I wasn't very fond of his arrogant, blunt self.

The next morning, I was up early and headed out in the snow for the hospital. Since this was a small hospital, there were only two beds in the Intensive Care Unit. Aunt Minnie was in the second bed, directly in front of the doorway. The entire area was very small, but well-staffed with four nurses. There was another woman in the other bed. She was alert and talking.

When I walked into the room, I went directly toward Aunt Minnie. I told the nurses that I was her niece. Aunt Minnie looked the same—tubes, IV's, swollen beyond recognition, and a pasty, ashy color. The ventilator made noise as it went on and off. I stood by the bed staring at her for several minutes. Seeing her in this condition was unreal. My prim and proper Aunt. Again this had to be a dream, and I wanted to awaken.

After a few minutes, two of the nurses came over and introduced themselves. I told them about my Aunt and how I couldn't believe this was happening. Tears flowed down my cheeks as I talked about this vibrant, feisty, proud lady. Both women smiled and listened patiently to me. They allowed me to unload, at least in a small way. I asked for a comb and brush, which they got for me, then I brushed Aunt Minnie's beautiful, soft gray hair and pulled it together. As bad as she looked, there was no need for her hair to be standing all over her head.

The nurse told me the name of the doctor assigned to Aunt Minnie. He was an internist, and of course they said he was very good. They expected him to come in around 1:30. I told them I would leave shortly for a while and then come back so I could meet with him. I turned and began to talk to my aunt. I leaned over talking directly into her ear. I remember telling her that she had to fight. I also reminded her of my promise to take care of her and that I would. I told her I loved her.

It seemed that within minutes the nurses were back in the cubicle and pulled the drapes. One said that they had to do a blood gases test on Aunt Minnie and that, if I wished, I could go to the lounge. I told them that I was fine and would stay with her. I was holding her hand. I had no idea what this test was or what it was for, so I asked. They said it was to determine the level of oxygen in her blood. Oh! Of course I still didn't understand.

First, they tried to stick these long needles in her arm. Her arm was three times its usual size. They talked among themselves, saying they didn't think this would work, because they couldn't find either a vein or an artery. One looked at the other and said, "We'll have to go in at the groin." The nurse reached for an instrument tray, picked up a scalpel and did what they called a cut-down. I was in complete shock, because as I stood there holding Aunt Minnie's left hand, she didn't so much as twitch. There was no movement at all, and these nurses had just cut into her to get this blood sample. It was unreal. It was like another bolt of lightning hitting me, making me face the fact that Aunt Minnie wasn't alive. This machine was breathing for her, her chest going in and out, her body warm, but she felt nothing. They had given no anesthetic at all, just cut her. I could have convinced myself they had hurt her if I hadn't been holding her hand. But I knew without a doubt that she hadn't moved. I also knew what I saw them do to her. In fact, I don't know if I can ever forget what they did to her.

Tears were running down my face, and I couldn't stop them. I wasn't crying or sobbing, but I couldn't control the tears. I stood there rubbing Aunt Minnie's hand and face and talking to her. My stomach felt like someone had kicked me in the gut. I felt weak. I felt panic. I was scared to death.

I told the nurses I would return. I remember one looked at me and asked if I was alright. As I left the hospital, I began crying as I walked through the corridors, across the lobby, and out into the parking lot. I got in my car and sobbed loudly. I knew then that my Aunt Minnie was really dead. At least, I believed she was, and I couldn't bear it. "No more! I can't take anymore!" I sobbed and cried out, "Please, Lord! Please help me! I can't lose her, too!" This just can't be happening. Aunt Minnie hadn't even been sick. I had talked to her Thursday night and promised to see her this weekend, but not like this! I was hysterical as I sat there sobbing in my car on that Saturday morning.

I picked up the car phone and called Marsha. She wasn't home, so I talked to Kristy. I told her how bad things were and that I was coming back to meet with the doctor later. I'd call again and let them know what he said. I called Daddy and told him I was just leaving the hospital and that there was no change in Aunt Minnie's condition. Although I had calmed myself, trying to be strong for him, evidently he could tell I was on the verge of cracking. Daddy talked very firmly to me about how I had done everything I could for Minnie. He was forceful in saying that we must have faith and put our trust in the Lord. He said I always wanted to control things and that I must realize there are some things that are out of man's hands. We must recognize and accept these things as the Will of God. I listened intently and knew he was right. This didn't mean he wasn't concerned about his sister. It meant he was concerned about me. With his words of wisdom, he was trying to pull me back so I could face the reality of what was happening. It was a heavy message, and I had heard every word my Daddy said.

After hanging up the phone, I sat there in the car silent. I guess I was trying to process all that was happening and all that had been said. I do believe I gained some strength that allowed me to pull myself together. At least for that moment.

I arrived back at the hospital around 1:15. The same nurses were on duty and welcomed me. There had been no change. The doctor had been in and was still in the building. They said he would be back to talk to me. Again I told Aunt Minnie that I was there, her niece, Carol Ann. I rubbed her hand and wiped her face with a cloth. The machine went on and off, breathing air into her body.

When the doctor arrived, he suggested we go down the hall to the small room that was used as a lounge. After introductions, he told me how bleak things looked. He was very kind and compassionate. He explained in great detail how grave things were for my aunt. He wanted to know about other family. I told him about my Dad and the recent, less than three weeks ago, loss of my mother. I told him that Dad hadn't asked to come to the hospital and that I didn't want him to see Aunt Minnie in this condition. I further explained that there were other nieces and a nephew. We were the only family, but I had been the one who had taken care of her since she went into the nursing home the previous year. The doctor looked at me and said, "This is my plan. If she lasts through the weekend, then on Monday, I'll have a neurologist consult, to determine if there is any brain activity. If that report isn't positive, then the family needs to consider removing the life support. Keeping her on life support beyond Monday could pose problems with euthanasia. I will need to know who will make this decision."

I was frozen in the chair! I looked up at him and weakly said, "I guess I'll be the one to make the decision." I was mortified! Never in my life would I have believed that I'd be in this position of saying "Turn off the ventilator." The doctor could see my upset and grief. He suggested I contact each of my cousins and let them know the situation. He said to tell them that if they wished to see Aunt Minnie, they should come to the hospital as soon as possible. There was no guarantee she would last even until Monday.

Although this was an unbelievably painful conversation, this man had listened to me, comforted me, and yet was honest and direct with me. We had never met before, but he extended himself to be of support to me and my family. He gave me his card and said I could call at any time or have the nurses page him. Reflecting on his kind manner, I thought he needed to go give that young doctor in emergency some lessons on interacting with patients' families.

I left the hospital and went to take Daddy his dinner. We sat in the kitchen, and I cried as I told him how serious things were. Daddy listened, and I know he was upset, but he continued to tell me to get myself together. I called our cousins: Lynne, Penny, and Rodney. I asked Lynne to tell her sister Michelle. I told each of them that if they wanted to see Aunt Minnie, they needed to go to the hospital. I also told them about her possibly being removed from life support on Monday. That night, I also called Marsha and told her. She was the only one out of town.

Saturday night, I went back to the hospital, and there was no change. The machine was just pumping on and off, on and off. The evening shift was as kind as the day shift. The nurses told me there had been several employees from the nursing home who had come by to see Aunt Minnie. Also that the nursing staff from there had called at least twice during their shift. One of the nurses looked at me and said, "Your Aunt must have been quite a lady." With tremendous pride, I said, "Yes, she is quite a lady, and everyone who knows her loves her!"

It was dark, freezing cold, and snowing as I made my way home. I kept thinking about how nice it was for the nursing home staff to show their concern. I know Aunt Minnie kept them on their toes with her quick wit and sharp tongue. She told it just like she saw it, and either you liked it or you didn't. But that was your problem, not hers.

twenty-nine

Around midday on Sunday, I was standing holding Aunt Minnie's hand when I looked up and saw my cousin, Lynne. She had come to see Aunt Minnie. I grabbed and hugged her. Of course, she was shaken at the sight of her in this condition. I asked her to walk down the hall with me to that little room. Lynne had lived with Aunt Minnie and Momma for a while, after she got out of college. Although she hadn't kept in touch with her, Aunt Minnie had helped her when she needed help.

Lynne and I sat and cried, then talked, then cried. I told her that the next day I would have to make a decision about the ventilator. I told her how I had come to believe that Aunt Minnie was already gone. Selfishly, I felt better that someone else in the family saw her in this condition, not just me. I knew without a doubt that my Aunt Minnie would not want to live like this. She was too vibrant, too full of energy, curiosity, and love of life to want to exist on a ventilator, in an unconscious state. Lynne, I think, agreed.

We went back into the room. Then we left the hospital together. When we got to the parking lot, I hugged her and thanked her for coming. I can't begin to describe how much this had meant to me, that someone else cared. I didn't feel as alone. Before Lynne went to her car, I told her that if the ventilator had to be turned off the next day, I would be there with Aunt Minnie.

Sunday night, there was no change. Aunt Minnie was the same. I combed her hair, wiped her face, and rubbed her hand. I told the nurses I wanted to talk to the doctor as soon as he got the report from the neurologist. I gave them my pager number, which they wrote in her chart with a note to the doctor.

That night I called Marsha again. We talked and cried for an hour. Marsha was so supportive throughout this whole ordeal. She was obviously really concerned about me. She was encouraging about how much Aunt Minnie had enjoyed the last year in the nursing home. She pointed out that this had been the best year for her in a long time and that we had to hold on to that. I can't begin to think of all the things we talked about, but I know I felt better. I was so grateful to have my sister and Gene to lean on. It was through them and the Grace of God that I was able to hang on.

Monday, February 5, I called work to say I wouldn't be in because of my Aunt's illness. Gene told me not to go to the hospital. I didn't respond. I knew I had to keep my promise. As soon as he left for work, I got dressed and went over to Daddy's. I got him up and told him we needed to go to the bank and close out Aunt Minnie's account and close her safe-deposit box. I explained that we didn't need to have any hassle, because she didn't have anything. He surprisingly agreed. Daddy's name had always been on her account since Uncle Billy died in 1976. My name had been added when she went into the nursing home. Daddy's name was on her safe-deposit box, and he had a key.

First we went to the bank where the safe-deposit box was.

Daddy signed, they pulled out the box, and I removed the contents and told the man we wanted to close out the box. Since I only had one key, they charged me $8.00 for the other key. Daddy was ready to start fussing about this being highway robbery. I told him to calm down. When we got in the car, I looked through the papers. The only thing in the box was Uncle Billy's divorce papers from his first wife and a certificate for two burial plots at Spring Grove Cemetery. I told Daddy what was there, and we were off to the next bank.

I had mailed the check to the nursing home on Thursday, February 2, so I knew there would be no more than about six hundred dollars in her account. I paid the nursing home $900+ the first of every month. Then every three months I paid $300+ to Blue Cross/Blue Shield for her medical insurance. This took all her monthly income, so whatever else she needed, I had paid for myself. I had never thought twice about the money I spent on Aunt Minnie, but on this day, I was truly grateful I had gotten her whatever she had asked for and more.

We closed out her account at Daddy's bank and transferred the balance into his checking account. Daddy had hung in there with me throughout this horrible experience. It was painful for me, and I know that it had to be for him, too. As a matter of fact, I was prepared to go alone, thinking he wouldn't come with me, but I was pleased that he did.

Just as I got into the car, in the bank parking lot, my pager went off. I looked at the number, and it was the hospital. I picked up the car phone, Daddy sitting there, and called the number. I was petrified. The nurse answered and said to hold on for the doctor. He began by saying he was sorry that the neurologist's report was as expected. There was no significant brain wave activity. I guess I was silent because the doctor repeated this and asked if I understood. I said that I did. He wanted to know my decision and if I had talked with other family members. I told him that I had and that my only request was that I be there when they removed Aunt Minnie from the ventilator. He said, "Certainly. If it's alright with you, I'll arrange it for 2:00 p.m." I said I would be there.

Daddy had heard the conversation and understood what was happening. I told him I couldn't let her die alone and that I had to do this. Daddy tried to say that I shouldn't put myself through this, but for once he wasn't forceful about it. I took him home. We sat in the kitchen and talked for a while. I told him I would call him from the hospital and that I was going to be alright. I told him, for the first time, that I truly believed Aunt Minnie had been dead since she slumped over on the couch Friday night. He made no comment.

I left Daddy's around 12:30 and headed back across town. Suddenly, I realized that I needed to drop off a case at the office. This was not out of my way and I have no idea how I remembered this. When I arrived at the office, most of the people were out to lunch. I rushed in and got the case ready to go out in the office mail. I was about to take off when I saw Judy, one of my coworkers. She looked at me and wanted to know what was wrong? "Is it your Dad?" I then told her about my Aunt. We tried to support each other as we talked about her mother and my Dad, Mom, and Aunt. She asked if Gene was going with me. I told her No, that he was at work and didn't even want me to go. She said, without any hesitation, "Then I'll go with you!" I was so touched. I just couldn't allow someone else to go through this, especially a friend. She was insistent, but I assured her that I would be fine. Judy gave me a big hug and a special blessing on that Monday afternoon. I will always remember her kindness.

I talked to myself all the way to the hospital. I arrived at 1:40 and went directly to Intensive Care. Aunt Minnie was the same as the night before, just turned on a different side. I spent a few minutes with her. The nurse asked me to go down the hall while they removed the ventilator. They would come and get me. It seemed like a long time, but I guess it was around fifteen minutes before the nurse came down the hall. As we walked back to Intensive Care, she explained that Aunt Minnie was only on oxygen now. That meant she would have to breathe on her own. There was no way of knowing how long she would survive off the ventilator.

When I went back into the cubicle, except for the oxygen tube in her nose, all the tubes were gone. I picked up Aunt Minnie's hand, leaned over, smoothed out her hair, rubbed her face, and bent down to talk to her. I told her that she was going to be with God and Momma. I told her that I loved her. I wanted her to go in peace. I told her it was no longer time to fight, but to go to The Lord. I told her I had kept my promise to her. I had taken care of her, and she wasn't alone. I told her I would take care of Daddy.

All of a sudden, the monitor started dropping. The alarm came on, and I turned around and saw the line go flat. I cried out to the nurse, "She's dying!" Both nurses came into the cubicle. One turned off the monitor, and the other put her arms around me saying, "She's gone." I stood there calling her name as if I didn't know what had happened. The nurses told me to take all the time I needed. I stood there looking at my dear Aunt Minnie, not understanding why she, too, had left me. She didn't last more than five minutes breathing on her own, if that long.

I walked over to the nurses' desk and sat down. One of the nurses got up, went in with Aunt Minnie, and pulled the curtain. The other nurse asked me for the name of the funeral home. She picked up the phone book, looked up the number, and called, telling them to come and pick up the body. I remember her being very firm about the pickup time, because, as she told the person on the phone, they didn't have a morgue.

After sitting there for a while, I began to feel like I was going to pass out. I asked the nurse if she would take my blood pressure. She did, and it was sky-high. She insisted they call my doctor and have a prescription called in to my pharmacy. They did this, and I promised I would go home and go to bed. I thanked them for all of their kindness and concern. I asked that they call the nursing home and notify them of Aunt Minnie's passing. They graciously agreed.

I called Daddy from the parking lot, told him that Aunt Minnie had passed and that I was with her. I told him I was going home to bed. I would call him later that evening. He understood. When I got home, I waited until four o'clock, then called Marsha. She would fly home on Wednesday. I notified the other cousins and Kim.

Here was yet another horrible experience that had fallen on me. I was totally and completely lost in my grief for Momma and Aunt Minnie. This whole thing was so unreal. There had been no indication of Aunt Minnie being ill. On Monday of that past week, we had talked on the phone. She told me she had a cold. On Thursday, when she called asking about her clean clothes, she said she was fine. Her cold was gone. We laughed and joked as usual. That's when I promised

I would be over during the weekend, no matter what. It made no sense.

But once again, looking back at that day, I do believe that despite Aunt Minnie's condition, there were:

Four Angels at Aunt Minnie's Bed,
Two at the Foot,
Two at the Head,
Four That Carried Her,
When She was Dead....

thirty

Both Marsha and Kim arrived on Wednesday. We had set the funeral for Saturday, February 11. This day would have been Aunt Minnie's 79th birthday. She had been so excited about her birthday coming. They made a big deal of birthdays at the nursing home. She told me about it every time we talked, because she wanted to make sure I didn't forget.

Now, here we were sitting around the dining room table with the funeral director, making arrangements, just as we had done the month before. It was horrible. Marsha once again pretty much took over handling things. I was just in and out of it most of the time. I did, however, once again, write the obituary. About the only thing I contributed during the meeting with the funeral director was that I wanted him to get that swelling down. I wanted Aunt Minnie to look like herself. He said he was pretty sure they could get most of it down. We gave him a great color picture of Aunt Minnie which was also used for the program.

Suddenly I realized there was a problem. Aunt Minnie had always told me that she bought three graves. One for Uncle Billy, one for his mother, and one for herself. Well, the papers in the lock box indicated only two graves had been purchased. This would mean that Aunt Minnie didn't have a grave, because Miss Jenny died after Uncle Billy. I was ready to panic until Gene said he'd find out. As it turned out, Aunt Minnie had told the truth. Evidently, she purchased two and then purchased the single grave. So there was a grave for her.

Something like this may seem minor, but when you're stressed out, all these little things make the situation worse. That's why I tell everyone now that they should have their business in order and spare their families additional problems and aggravation.

This whole experience was like *deja vu*. Here we were again, in the living room talking with the minister. Marsha again stepped up to the plate and told him all about Aunt Minnie. And once again, I sat silent, unable to deal with this discussion. Daddy also just sat there. We both were like zombies.

Kim pitched in and answered the phone and kept track of things that people dropped off. She also was my support, along with Gene, when I would come home emotionally and physically spent.

I decided I needed to buy Aunt Minnie something to wear, because there was no way her clothes would fit her. I was not up to shopping, but had no choice. Finally, I settled on a gray coat dress. It opened down the front, but had a low V-neck. I told the sales woman why I needed the dress. We both decided I could get a pretty scarf for around the neck. So I bought this outfit and a scarf. When I got back to Daddy's, Marsha found one of Aunt Minnie's scarves that she liked better, so we used that one. It hadn't crossed my mind that Aunt Minnie had about fifty scarves washed, ironed, and folded neatly in a box in the basement. All of this had been done by Momma.

Mrs. Harris, a member of Aunt Minnie's Circle 9, called

and said they wanted to prepare dinner for the family after the service. I told her that we really appreciated it, but I didn't think we would have anything. Later I talked to her again, and she said the Circle would have continental breakfast, downstairs at the church, before the service. This was very nice and thoughtful. She felt people might enjoy coffee and juice, especially since the weather was so very cold. She and Mrs. Taylor arranged this, and we were all most appreciative. These were the two faithful women from our church who had visited Aunt Minnie in the nursing home.

The morning of the funeral, the temperature was sub-zero. Although the roads were clear, the ground was covered with ice and snow. Lynne was the only cousin who came to the house to ride with us in the funeral car. That meant it was Daddy, Marsha, Gene, Kim, Lynne, and myself.

When we arrived, the men were still setting up the casket. There were a few floral arrangements on each side. As I walked down the aisle toward the casket, I felt cold chills. I looked at Aunt Minnie in that coffin, and she looked terrible. There was no way anyone could even recognize her. She was as bloated as before her death, still ashy colored and her features unrecognizable. I went over and sat down crying. I think maybe it was Marsha who said, "Don't you think we should close the coffin?" I said yes. I remember turning to Lynne and asking if she minded. She said she didn't, so I went up and told the funeral director to close the casket. I probably told him she looked terrible, because I remember him trying to explain they had done the best they could, that she had been on life support for several days and this does something. But I also remember it looked like they had glued her lips together. Anyway, the casket was closed. I will never regret this decision. Aunt Minnie was very prim and proper. Her appearance was important to her. There was no way she would have wanted people to see her in this state.

The service was very nice, and once again, Mrs. Robinson read a beautiful tribute to my Aunt from the church. As I sat in the front row, I kept staring at a plant that was on the far side of the casket. It looked like a twig, but had this large blossom. Later, I found out it was an orchid sent by the Edwards family. Aunt Minnie was the baby-sitter for their daughter for several years. They knew that she grew orchids and wanted me to have this one. This was very touching.

All of Daddy's family was there: Rodney and Jimmie, Penny and Sam, Lynne, Michelle and Ron, Lloyd, Mallie, Willie and Lisa, Roger and their children from Michigan. Despite the terrible weather, there were a number of her friends who came out to pay their respects. On this day, her birthday, Aunt Minnie would have been pleased with the service.

Aunt Minnie was buried in Spring Grove on a hillside overlook, next to Uncle Billy. It's a beautiful area, although on this day, the wind-chill was well below zero and that wind was whipping across the hillside something terrible. The grave site service was short, and there was no hanging around. Everyone was scurrying to get back in their cars. That was it. This was her birthday, and this was her burial day.

Daddy held up through all of this as if nothing had happened. He completely internalized his feelings and talked as if it was someone else's sister. Marsha and I were concerned about this and didn't feel that it was healthy. However, on several occasions, Daddy did sort of lecture me, when he saw I was near a breaking point. It was not fussing, as he usually did, but more trying to help me accept what had happened. I believe now that Daddy felt he had to be strong for me. He had to help me keep going. I love him for this, and his words did help me.

Sunday night, both Marsha and Kim flew home. I had to carry on and do all the things that have to be done when death occurs. I was right back where I had been a month earlier. The nursing home called the house about Aunt Minnie's belongings. I asked that they secure them until I could get them. I told Gene there was no way I could go over there, so he handled that for me. They had placed all her things in large plastic bags. He put them in the basement storage area and told me to leave them alone until I felt like going through them. I washed the soiled clothing and put it down there also. To this day, I wouldn't know if anything was missing, because I couldn't face going through her things for over a year.

thirty-one

My loss of Aunt Minnie was great, but my loss of Momma was all-consuming. I thought about Aunt Minnie, missed her, and was very upset about losing her. But I was devastated without Momma. I'd go to the cemetery several times a week to visit her grave. Sometimes it was very peaceful, beautiful, and serene. Often I just sat in my car thinking and talking to her, telling her what was going on and how I felt. Other times, I'd go and cry, uncontrollably. There was no way to know what each visit might bring. There still isn't. If I'm on that side of town, I often eat lunch there, sometimes in tears, sometimes not. I know my mother and Aunt Minnie are in heaven. Of this, I have no doubt. However, I cannot get over needing them or having lost the connection and sharing we had. This was so powerful! I didn't realize it until they were gone.

Family is so important, and bonding with family is even more important. I know I must have done something right in this lifetime of mine, because I have truly bonded with my daughter Kim, the same as my mother and I. This is a blessing I pray will last my lifetime. I treasure every minute of our time together and our conversations, both serious and light. Kim has always been the joy of my life and knowing that she loves her mother, makes her even more special.

The death of my mother caused another change in my life. It strengthened the bond between myself and Marsha. Marsha had lived away from home for almost thirty years. There certainly wasn't ever a problem between us, but in retrospect, we weren't really close. When she came home several times a year, she stayed with Momma and Daddy. She called them every Saturday. I talked with Marsha maybe once a month. When she visited, I was always over there, but it was she and Momma who did things together. I felt it was *their* time together. Consequently, it was never Marsha and I doing things.

Strange as it might seem, out of Momma's death came this connection or bonding, as I call it, with Marsha. We had to support each other, lean on each other, use each other's strengths, and find common ground just to survive the unfolding of each new tragedy. We had to communicate our feelings, we had to give each other space, and we had to work together to support our father. Our blessings have been abundant and our tears have flowed, but our ties have been strengthened, through faith and trust in The Lord.

I often sing the gospel song, "Hold on, Help Is on the Way." This is what each of us in our family has tried to do, "Hold On," for the blessings of one can spread and become the blessings of many. We know that our mother was a "Child of God." We know that Aunt Minnie was a "Child of God." Our job now is to make sure that we, as well as our children and grandchildren, continue to live righteous and worthy lives.

I returned to work the Wednesday following Aunt Minnie's funeral. Of course, this was additional pressure because I had to get my work caught up—again. The staff had been helpful, and I was appreciative. But I remember going in that morning and facing sixteen voice mail messages. That means sixteen problems. So I had to dive in and sort things out as quickly as possible.

I no longer had the nursing home and laundry on my schedule, but I still had to go and check on Daddy. He just outright refused to consider living with Gene and myself or at an Assisted Living Facility. He went completely off any time this was mentioned. There was no reasoning with him. He even refused to go visit Marsha or Kim. I mean not even visit! He didn't want to leave the house other than to go to church, the cemetery, or if necessary, the eye doctor. Daddy would not come out to our house even for Sunday dinner. He acted like we were going to force him to stay.

He did start to get on me about doing his income taxes, though. This was late, usually starting on me the first of January. He wanted them done every year when the forms arrived. This year had been an exception. Not only did I do his taxes, but also Aunt Minnie's, Kim and Toren's, and our own. This was all on top of trying to get legal affairs straightened out for Momma. Now I had to start all over and do the same for Aunt Minnie. It was a huge job, and Daddy felt better once this was all taken care of. I realized that he sat there all day thinking of things that needed to be done, but I was worn out. He seemed to think I could fix things instantly. I tried my best to do all the things he wanted, to keep him from worrying. Sometimes this was possible, and sometimes it wasn't. I just did my best to keep both of us going.

This period was hard on all of us, I'm sure, but I knew I was barely functioning. I had never experienced anything like this. I went through each day wearing a mask, acting as if things were under control. I had splitting headaches that started in late afternoon and continued to get worse. So every evening when I got home, not only was I exhausted, but I had a severe headache. I would go straight to bed. Sometimes I would eat and sometimes not.

It's hard to explain, but there was this feeling of being out of control. I couldn't fix things. I couldn't make things better for those I cared about. I was so deeply entrenched in my grief that I couldn't focus. This wasn't natural for me, not since I was the one who usually stayed on top of things. It's difficult to describe, but this was all going on inside me every waking minute. I was suddenly and totally disorganized, couldn't remember things, and was simply overwhelmed.

By the end of March, I decided, with prompting from Gene, to go to the doctor. Since we had changed insurance companies, it was necessary for me to find a new physician. This was the worst possible time to have to change doctors, because I'd have to go to someone who didn't know me—another thing I didn't have any control over. I found a doctor close to my home.

On the day I went into his office, after work, I was without a doubt a walking zombie. I had decided I needed help and that I would have to explain what had been going on in my life. How could I do this with a stranger? He would think I was nuts! How could I put into words what needed to be said? All this was going through my mind while I sat in the waiting room.

As each year in my life has passed, I've become increasingly spiritual. I have come to rely on "The Lord" more and more. I have been able to look back at my life and accomplishments and recognize "His Hand" in things. I have stepped out on "Faith," as well as triumphed in victory through "His Grace." Despite my despair, I never for one minute felt forsaken. I knew deep down in my soul that I had to continue to give praise. I knew that "The Lord" had not forsaken me or my family, but I needed help!

As I sat there in the examining room reflecting on these feelings, I began to cry. I started talking to myself, saying I had to get myself together before the doctor came in. After all, I didn't even know this person. I knew I was in bad shape, but I didn't want to be committed. So I pulled myself together, or at least that's what I thought.

When the doctor walked into the room, he greeted me, pleasantly, extended his hand, and sat down at the desk. He had read the nurse's notations on my chart before coming into the room. It wasn't surprising when he said my blood pressure was high. He checked it again himself, and it was still up there. His manner was kind, and he was soft-spoken. I immediately felt at ease. He said very directly, "Tell me what's going on with you."

Without hesitation, I unloaded my whole story of the past seven or eight months. I started with the joy and excitement surrounding Kim and Toren's wedding. I continued with every detail of the events following. I talked and cried and shared my inner pain and loss. I remember describing the guilt I felt about Aunt Minnie's death. I loved her and missed her, but my loss of Momma was consuming, my thoughts of her constant. As a result, I felt I hadn't been able to give Aunt Minnie her due respect. I said I knew I had kept my promise to Aunt Minnie, to take care of her, but I still had this feeling of guilt. Or at least I couldn't think of another word to describe my feelings.

This doctor sat listening and not for one instant did he seem impatient. He allowed me to unload, to cry, to explain and just share, and he was a stranger. I shall always be grateful for his time and compassion. Without any hesitation, he said I was depressed. He also said I was going to have to get over being so hard on myself. His concerns were for getting my pressure under control, getting a blood work-up, and assuring me that I wasn't crazy. He changed my blood pressure medication and prescribed an anti-depressant, saying it would take a couple of weeks to make a difference. My problems weren't going to disappear, but he felt strongly that this medication would help me handle difficult situations better. He was very supportive in talking with me about the things that had happened. He felt that, despite my expressed feelings of guilt and helplessness, I was stronger than I gave myself credit for. His parting words, after telling me to come back in two weeks, were that I needed to take more time for myself.

On this day, in this office, I had unleashed my soul to a stranger. He had listened. He had not judged but had offered comfort. I liked this person. I felt his compassion. I knew I had done the right thing, and I was thankful to God for leading me to someone who cared.

thirty-two

Marsha took time off work and flew home for a couple of long weekends. This meant a lot to both Daddy and myself. Usually she came in on Wednesday evening and would leave Sunday evening. During each trip, she spent her entire time working around the house. This also allowed us time to do some things together. She had the use of Momma's car while she was here. Daddy was pleased to have someone drive the Buick Century. He insisted it remain in the garage, so I had to buy new license plates as well as pay the insurance. Several times I tried to convince him to give it to Marsha. Each time he had an excuse. He just needed that car sitting in the garage.

I continued talking with the Hospice Home Care counselor, too. She was another source of inspiration for me. Often she called me at work. This gave me the opportunity to discuss how concerned I was about Daddy. He didn't have any business sitting in that house 24 hours a day by himself. I told her how this was a constant worry. Even with my going over every day, there were numerous times when he didn't answer the phone. Or the phone would be off the hook. This meant pure panic for me, Gene and I getting in the car and racing eighteen miles to his house. Each time I was in total panic, imagining the worst.

When we'd arrive, either Daddy was asleep or just sitting there, not knowing the receiver was off the hook. This upset me to no end. It was difficult to calm myself down after one of these incidents. Daddy would start fussing about us coming over. This added to my irritation. He would say something like, "If I am dead, then I am just dead!" Now this really helps when you're upset. He doesn't seem to understand how upsetting it is to get no answer after letting the phone ring 25 times. He'd just say, "That's your fault for letting it ring that long!"

Having shared this with the counselor from Hospice one day, she said very directly to me: "Your father will not leave that house. He is willing himself to die there. This is his wish! You must accept that you cannot change this!" I was speechless for a minute. I fully understood what she was saying, but I had trouble accepting this. I told her I couldn't handle walking into the house and finding Daddy dead! There was no way I could handle that! No way!

I must have been pretty emphatic, because she didn't pursue this any further. The counselor had visited Daddy several times. She discussed those visits with me, but always in general terms. I respected her for this, because she didn't betray his trust. Not, that is, if he shared any feelings with her. This we will never know.

One of the things Marsha and I talked about often was how there was such a tremendous outpouring of love, support, and caring when someone dies. This lasts for a brief period, until after the funeral. Then all these people move on, back into their daily lives, while the grieving family is left alone. It's as if there's an expectation that it's over and we should carry on. The visits stop, the phone calls stop, the support from others just stops! All the people who checked on Momma, both family and friends, didn't do the same for Daddy. This was hurtful for him. He was definitely aware that since Momma's death his existence has been overlooked by many. He will sit and tell me how he hasn't heard from this person or that person. He will go back and tell me all the things that both he and Momma had done for them. He remembers every detail. This makes me feel so sad. It's something else I cannot seem to fix.

Marsha is right about how Daddy became bitter when he lost his sight. He would say sharp, cutting things to people. This is true, and he spared no one with his criticism, including Momma, Marsha, the kids, and myself. However, this wasn't the father we grew up with, nor the person Momma married. Throughout our childhood and growing up, I can honestly say that I don't remember Daddy and Momma ever arguing or fighting. Daddy was always firm but had a pleasant nature and would help anyone. He worked hard, was thrifty, was caring, and loved his family. He imbued in each of us his love of The Lord. He worked side by side with Momma in the church until he became disabled. I personally know of many situations where both Momma and Daddy helped members of his family and her family, by lending them money, and their home was always a haven for those who needed a place to stay.

These are the things my father remembers. He doesn't realize how the change in his personality pushed people away from him. He doesn't see that he is extremely critical of others. He doesn't realize that people have come to avoid him because of his opinions. He does recognize that the people he helped are no longer there for him. This causes him great pain. This causes me great pain.

Whenever he brings this up in a conversation, I tell him the same thing. I try to explain that we are the ones who were closest to Momma and Aunt Minnie. Our grief is long-lasting, whereas others expressed their sympathy then went about their day-to-day lives. This is the same way we have done when others died. Now, I thought this sounded good. It is certainly true, but Daddy never seems to buy this as an excuse for people ignoring him. Again, this is something I've not been able to fix for him. I wish I could.

Each year when spring arrived, Daddy's mood became more positive. This was because he could get out in the yard and cut the grass. Of course, he had no business doing this, but it was one thing he looked forward to. It worried us to death, as well as the neighbors, because he had so little vision. His yard is on a corner lot with lots and lots of grass. The only good part is that all but a small strip is flat. So, every day, Daddy gets out his Toro and cuts grass. It takes him almost the entire week to get around the house. Then it's time to start all over. We know he enjoys this. It does give him something he can do, and not only is it important to him, it's an accomplishment. But he works so long each day that he winds up totally exhausted, and that scares us. So both Marsha and I tried to talk him into having it cut.

When Marsha came home in the spring, she and Michael cleaned out the flower beds. They put down mulch and planted flowers. These were things Momma did in the yard every year. Marsha also put out hanging baskets on the front porch, then filled the planter on the side patio with geraniums. Daddy really liked this, because things were just as if Momma had done them. The only difference this year would be that Daddy was giving up his garden and his strawberry patch. He said he could no longer fool with them. There would no longer be fresh homemade strawberry jam made by Momma.

thirty-three

By late spring, I felt that the anti-depressant was helping but found it difficult to describe how. The only thing I could pinpoint was that I wasn't crying as much. The doctor was encouraging. He attributed my headaches to stress. He indicated that I must continue to work on coping with things I couldn't control. I knew this was a problem. It always had been.

Gene planned an August trip for us. We were going to take a 14-day cruise through Scandinavia. I was thrilled, but I couldn't help feeling uneasy about leaving Daddy, even though Marsha would be here while we were away. This trip was something positive and exciting on which I could focus. Daddy was pleased and enjoyed telling me about all the countries he and Momma had visited. He even told me, "You should travel when you're young!" This was a real shock—Daddy not lecturing me about spending money.

However, Mother's Day was approaching, and I was dreading this. I couldn't comprehend that my mother was dead. I continued to go to the cemetery regularly. I always found it to be helpful, no matter what my mood. I sat there one afternoon in my car asking for strength to face Mother's Day. I also talked some with Marsha, and our feelings were the same. We just didn't want to face this day. It was like having an open wound, then someone cuts deeper into it with a knife. The hurt was bad.

On Mother's Day morning, the phone rang early. It was Kim and Toren calling to wish me Happy Mother's Day. We talked almost daily, and I had been open with Kim about my feelings. She was always comforting when I was depressed. On this Mother's Day, 1995, Kim and Toren told Gene and me that they were expecting a baby. I went into orbit! We were both thrilled! Although I had four step-grandchildren and love them dearly, this would be my first biological grandchild. While I was going on and on, Kim broke in and said, "Momma, my due date is January 10th." I couldn't say a word then. I didn't know if Kim knew this was the date Momma died. Later, I found out that she did know, but we didn't discuss it on that Mother's Day.

The Lord brought our family this wonderful news on a day when each of us needed an uplift. It reinforced our many blessings, despite our grief and pain. It gave each of us a new focus. Deep down in my soul, I believed that The Lord was giving life back to our family. He was letting us know that, Yes, we were loved. Despite ourselves, despite others, our family was truly loved. I reflect on this day with praise and thanksgiving because this news brought joy to our hearts. How could one ask for a better Mother's Day present? It wasn't just a present for me, either, but a shared present with Gene, Daddy, Marsha, her children, and our entire family. A Blessing from Above!

Daddy was thrilled! He sat all day worrying about Kim eating right and taking care of herself. Kim has always been very active. Getting her to slow down was going to be difficult. We all tried to talk to her, but her response was always,

"I'm fine!" Both she and Toren decided they didn't want to know the baby's sex. I was surprised, because Kim is one who cannot keep a secret. This was fine with me. It just meant we all had to wait for the birth.

Then, during the early part of June, Marsha got sick and had to have emergency surgery. This was another trauma for all of us. I felt terrible because I wasn't able to go up and stay with her. First, Daddy didn't want to leave home. He felt he couldn't help and that he'd be in the way. He told me to just go on and he would be fine. Well, that wasn't true, and I couldn't leave him. But, second, I had been off so much from work and had already scheduled the three weeks in August for our trip.

Again I felt guilty, because I wasn't able to be there for my sister during her illness. The best I could do was to call her daily. Marsha was very understanding about my not being able to come to Maryland. I knew that if Momma were alive, she would have been at her side. Marsha was going to have to recuperate for about six to eight weeks after the surgery. This was the pits, because she would lose her entire summer vacation. However, we were blessed that all went well with the surgery.

Marsha's illness meant I had to make other arrangements for Daddy in August. Marsha and I finally worked out this plan. Michael would fly to St. Louis and stay with Kim and Toren for a few days. Then Kim and Michael would drive here and stay with Daddy. That way Kim wouldn't be on the road alone, since she was expecting. Although Michael didn't have his license, at least Kim would have company on the six-hour trip. These arrangements met Daddy's approval. He was delighted to have the kids coming. It was settled.

thirty-four

The summer went quickly. My schedule was still the same. Go to work, go to Daddy's, come home and go to bed. I felt a tiredness that wouldn't go away. I was dragging and sleep didn't help. But when duty calls, we must go. So, each day, I found myself dragging out of bed and starting my routine over again.

There was still no persuading Daddy to go to one of the Senior Day Care Programs or to do anything other than stay at home. Lu came each week, so the house was well-maintained. Every day, I cleaned up pipe tobacco off the table and the kitchen floor. Sometimes I'd go to the bedroom and say out loud, "Momma, I understand now how irritating it is to always be cleaning up tobacco." This would make me laugh! It was like a private joke between Momma and me. I was the one who used to tell her not to let it get on her nerves. Now it was my turn—except I only cleaned up once a day, while she had to do it morning, noon, and night. I finally understood.

It was around this time that I realized I consciously wasn't going to church every Sunday. Daddy and I always went on the first Sunday, Communion Sunday. Other Sundays he was not interested in attending. Sometimes I would go without him, but if he wasn't going, I had an excuse to stay in bed. I began to realize that Sundays were depressing for me.

If I went to church, it reopened the wound that Momma wasn't there sitting on the pew with me. Each service became an emotional experience for me, thinking of Momma. Sometimes this meant tears and sometimes not, but I always left with a heavy heart. After church, I would go to the cemetery. This, in itself, wasn't a problem, but remembering Sundays and the role Momma played in our lives was.

She would always come home from church, put on her apron, and fix dinner. She was usually rushing because she had either a Wilberforce Alumni meeting or a Circle Six meeting at four o'clock. Most of the time, I never stayed for dinner, just sat there in the kitchen talking to her as she scurried around. She had to get dinner on the table, and the kitchen cleaned up before she left. Daddy would be watching a game. On the Sundays that she didn't have meetings, Momma and Daddy would come out to our house in the afternoon. They always stayed until after *60 Minutes*. This was the routine. I missed this terribly! I still have trouble with Sundays.

Before I knew it, we were about to leave on our trip. I had been extremely tense about going out of the country, leaving Daddy. Since the cruise included a port in Russia, I had to get a tetanus shot and a diphtheria shot. While I was there, the doctor gave me a good talking-to about how much I needed this trip. I knew he was right, but I felt uptight about leaving. We had even bought trip insurance to be on the safe side, since things had been so rocky.

As I drove home from the doctor, I suddenly thought about Gene. He was ecstatic about this trip. Then, like a lightning bolt, it hit me: I owed him this trip! Gene had been there every step of the way, for Kim, Daddy, my family, and me. He had put up with my being depressed, irritable, sick, and everything else. Gene had been my rock through all of this "stuff," and I needed to think about him for a change. Most people would complain about all the time I spent looking after Daddy, but not Gene. No matter what I asked him to do for Daddy, he would take care of it. I thought of the number of times we had rushed over there because I couldn't reach Daddy. All these things I had been taking for granted. Yes, I needed a break, but what was more important: Gene needed a break.

The ride home from the doctor allowed me to feel a sense of relief about the trip. It was almost as if a Higher Power was sanctioning my taking this break. Suddenly, I was excited and ready to go on what would be the trip of our lives.

Fortunately, by the time we left, Marsha was up and around. Although she couldn't come to stay with Daddy, she was recuperating well. I gave Daddy all his instructions about not being too hard on the kids. He just laughed at me! It seemed every other question from him was about when we'd be back. One day had become like the next for him, so it was difficult for him to remember days and dates.

On August 15, Gene and I flew to London. We boarded a bus, drove south through the English countryside to the port at Harwick. As we approached the pier, there was our ship, *The Song of Norway*. We were on our way! I felt good. I felt excited. I felt a quiet peacefulness. I was glad we were there.

However, once we got to our cabin, I told Gene, "I hope we haven't spent all this money on a Senior Citizen Cruise." We both laughed. It seemed that everyone we saw had white hair, a cane, or just looked old. Now, that's terrible to say, but it's true. We cracked up about this as we unpacked. In fact, Gene took a video of me acting crazy in the cabin. It had been a long time since I'd felt so light-hearted and carefree.

At dinner that evening we met our table mates. We were seated with a family of five from Mexico City: Tom, Marcia, and their sons, Tom Jr., Andrew, and Mark. The boys were about 13, 10, and 8. They all spoke English fluently, so there was no language barrier. This family made our trip exciting each and every day. Gene and I enjoyed each meal. The boys kept us hysterical.

Our ports were: Stockholm, Helsinki, St. Petersburg, Tallin, Copenhagen, and Oslo. Each of these cities left us with special memories which will remain with us for a lifetime. But the trip's ultimate experience, for me, I owe to Tom Jr. At breakfast one morning he told us about getting up and watching the sunrise from the upper deck. He kept telling me that I should do this. I laughed. There was no way I could wake up that early.

The next morning, for some unknown reason, I awakened and looked out the porthole over my bed. It was dark, but there was a glimmer of light on the horizon. I got up, put my slacks and top on over my gown, picked up Gene's camera, and went up on deck. There in the dark, I stood alone, listening to the ship move through the water. I could see a strip of light on the horizon. It was windy and cold. I tried to focus the camera but was doing something wrong.

After fiddling with it for a few minutes, I rushed back to the cabin. I awakened Gene and asked him what I was doing wrong with the camera. He thought I was out of my mind, but he did something to the camera. I grabbed my windbreaker, put it on, and headed back up on deck. Once again, there was no one there but me.

I stood at the bow of that ship on that morning and

watched the sun slowly rise over the horizon. It was the most majestic experience I've ever known. I lifted both my hands to the heavens. I prayed out loudly to God for help and strength. I prayed for Momma, Daddy, and Aunt Minnie. I prayed for my unborn grandchild. I prayed for our souls. I felt the presence of The Lord.

That morning, on the bow of the ship, I experienced the true beauty of God's Kingdom here on earth. I saw the sunrise on the Baltic Sea, as we headed for port in Helsinki. I still give thanks for that opportunity to commune with God. The experience will be with me always. Truly the blessings of The Father are real and bountiful. Our trip was a blessing for us both.

thirty-five

Returning home, I found that everyone had survived without me. I realized I hadn't had a headache the entire time we were away. This proved the doctor's point about stress. Daddy was delighted to have us back. He enjoyed our stories but got greater satisfaction from telling of his own travels with Momma. That was alright—he was happy!

I immediately went back to work. My daily routine was back on track. I had one large task facing me: to file estate taxes for Momma. This was due in October. The forms were very complicated, but I was determined to do this and not hire an attorney. So I waded knee-deep through paperwork. I had to go back and forth to the courthouse and county offices getting assistance. I got everything filed by October 1st and felt great satisfaction over accomplishing this.

Kim was doing fine, although she hadn't slowed down and her work schedule was hectic. Kim is an educational consultant on HIV/AIDS, doing seminars for universities and colleges and forming awareness groups. Now, this daughter of ours had submitted a proposal to the National Alliance of Black School Educators to do a seminar at the annual meeting in Dallas in November. She had been upset the year before when she looked at our program and found there hadn't been any sessions on HIV/AIDS. I knew how she went off about this, but I didn't know she had followed up. Well, she was scheduled to present at the Dallas Conference—and now would be seven months pregnant. I was concerned about her doing too much, but her doctor said she could go and her mind was already made up. So that was that.

When Gene and I went to the convention in November, we hadn't seen Kim for a while, so she looked gigantic to us. She looked beautiful, too, and as vibrant as ever. We had a great time spending that day with her. What I remember most was how much she ate. I was amazed at all the food she ordered. I told her she was going to look like a walrus before the baby came, and we just laughed. The baby was kicking like crazy. You could see her stomach moving. Our grandchild was making his or her presence known.

Gene and I were both in awe at Kim's presentation the next morning. Our daughter was dynamic! I knew she had a wealth of information about HIV/AIDS, but we had never seen her present. We were so proud of her poise, grace, style, and knowledge. We were ready to pop! She kept people's attention for two hours and fielded questions with ease. It was an excellent presentation.

When Gene and I came home on Sunday, we found that Daddy had done just fine. Again, he was glad to see us. He wanted to know all about Kim. He, too, was proud of her.

The following week was Thanksgiving. Marsha came with all her family. We pulled together a dinner, which everyone seemed to enjoy. It was not special like Momma's, but it served the purpose. Marsha told me during this visit that there was no way she could come home for Christmas. She felt it would be too painful and that she would probably stay in Maryland with the kids. She also felt we should try to start a different tradition, because it just couldn't be the same without Momma. She also thought maybe Gene and I needed to be with Kim and Toren.

I understood how Marsha felt. Until she brought it up, I hadn't said anything about Christmas. I was dreading Christmas! I had been worrying about it for some time. I knew that Kim wouldn't be able to travel. I knew it was going to be a painful time for all of us. I knew I couldn't face trying to have the Holiday Open House without Momma. I knew Daddy was going to have a rough time. I just wanted it all to go away. I didn't want to have to pretend that everything was alright!

Finally, I told Marsha that we should just play it by ear. If she stayed home, then we would do something with Daddy. Daddy had made it clear that he wouldn't go to Maryland, so that wasn't an option. He wanted to stay in his own house.

Right after Thanksgiving, Kim began having problems with the pregnancy. She was in the hospital for a couple of days, then on bed rest. The baby had fractured her rib, kicking so hard. Her blood pressure was spiking, and the doctor was concerned. She had to stop work, no more running around to all these universities. Her office was great, buying her a laptop computer with a fax modem so she could work from home. Not only did this help them, but it helped her keep busy. Kim has never been a complainer, so she insisted we didn't need to come to St. Louis. That child of mine is a mess! She has always been independent, although this never stopped me from telling her what I thought.

Mid-December, Kim went to see the doctor, and she had pre-eclampsia. He wanted her to go straight to the hospital. So, Kim was admitted. Toren was a nervous wreck, but he assured us that Kim was doing fine. The doctor was running some tests. The baby's vital signs were strong. Kim didn't want her Dad and I to come. In fact, she insisted, saying that she only wanted Toren there, that she wasn't putting on a side show. As much as I wanted to get on a plane, Gene convinced me that I had to honor her wishes.

By Saturday morning, I was a nervous wreck. The baby was due on January 10, and this was December 16.

It seemed as if every time I called the hospital Kim's line was busy. Now, here we are 400 miles away, trying to find out what's going on, and she's chatting on the phone. Not only that, but by this time I had Daddy, Marsha, Heather, and everybody calling me to find out what was happening. I finally called the nurses' station. I told her that I was Kim's mother and asked if she would go in and tell her to call me. The nurse was very kind. In a few minutes, the phone rang. It was Kim. Just as I was ready to jump on her, she started laughing. She said Toren had fallen asleep in the chair after talking on the phone. It had been off the hook.

Early that morning, the doctor decided to induce labor. Kim said he felt they needed to take the baby now. She tried to tell me all the details of the process, every step, over the phone. I was on the other end of the line saying "What?" every few minutes. Kim was as calm as could be.

So, on Sunday, December 17, 1995, Kim and Toren made us the proud grandparents of Brandon Luke Jones, weighing in at 6 pounds 12 ounces and 21 inches long. A healthy, beautiful, blessed baby boy joined our family on this Sunday afternoon. Both mother and baby were fine. The father and both sets of grandparents were ecstatic. We shared the news with anyone who would listen! God is Good!

thirty-six

The birth of our new grandson lifted a cloud that had hung over our family for what seemed ages. The news of Brandon's birth spread quickly throughout the family. It changed our entire focus from dreading the upcoming holidays, to rejoicing, giving thanks to The Lord for the arrival of this healthy child. Our grief over the loss of Momma and Aunt Minnie wasn't gone, but this special blessing helped us look beyond the grief.

As we shared the news with others, almost without exception, those close to us would comment how proud Harriet would have been! Without hesitation, I would answer that Brandon was blessed with two guardian angels, Momma and Aunt Minnie.

It felt perfect that when I got the phone call about Brandon's birth I was with Daddy, sitting at the kitchen table. We had been discussing Kim and how long she had been in labor. The phone rang, and Gene belted out, "It's a boy!" That's all he knew from Toren, that it was a boy, and that both Kim and the baby were fine. Toren, according to Gene, was all out of breath and said he had to get back in the room. It was later that evening before we got all the details and the baby's name. We knew that the middle name for a boy would be Luke, after Toren's father. They kept changing their minds about the first name. Anyway, Brandon Luke Jones, entered this world with lots of people to love and cherish him.

I was so happy to be there to share this happy time with Daddy. He was thrilled, and it was the first time in over a year that he had something to celebrate. Yes, it was exciting when we found out that Kim was expecting, but this was the best! Daddy had another great-grandson! And he didn't have to hear the news alone—I was there to help him enjoy this event. It meant a lot to both of us.

Marsha and I called back and forth every day now. She definitely was not coming home for Christmas. Gene and I would be going to St. Louis. I told Marsha that we had to pray Daddy would go also. I was afraid to bring it up. If he refused to go, then I couldn't leave him here alone. Well, when I finally got my nerve up, I didn't ask, I just told him that we had to go to St. Louis to help Kim. By putting it this way, he didn't get a chance to refuse. I told him she needed help with the baby and that she was exhausted. Daddy looked up at me and said, "You'll have to pack my clothes." Another blessing!

Of course, before we left town, I had a million things to do. I made a mad dash to the stores to buy more baby things. The most important thing was that I couldn't leave town without getting Christmas wreaths for Momma's and Aunt Minnie's graves. The ground was covered with about three inches of freshly fallen snow. I got in the car and put my boots and a broom in the trunk. I stopped by the florist, had two beautiful wreaths made out of evergreen, boxwood, pine cones, and a large red bow. I was really pushing it to get to the cemetery before closing. There was no way I could leave town without remembering Momma and Aunt Minnie.

Spring Grove Cemetery is the second largest cemetery in the country, and Momma is buried in its Garden Sanctuary, right off the main entrance. The air was cold as I pulled up and put my boots on. I took the broom and wreath and had no trouble finding Momma's grave plaque in the snow. I swept it off and placed the wreath on top. As I walked away, I turned to see this lovely green spray with the big red bow sitting amidst the snow. Momma would like this, I was sure.

Now, Aunt Minnie's grave site was on the far side of the cemetery. There was a sign saying, "Roads Dangerous, Travel At Your Own Risk." So I went out the main gate and followed the city streets to the back entrance. I had no trouble getting to the hillside where Aunt Minnie and Uncle Billy were buried. As I got the wreath and broom out of the trunk, I suddenly realized that finding Aunt Minnie's grave was going to be a major feat. I'm pretty good about directions, but looking across that hillside, there was nothing but a blanket of white. I walked across the hill, making footprints in the fresh snow. I decided I had to be close, so I put down the wreath and started sweeping, trying to find her bronze plaque. This went on for maybe fifteen minutes before I stopped. I had moved several times to different areas, but with no luck. Finally, I looked around this snow-covered hillside and felt it was hopeless. There wasn't a soul—I guess I'd better say "a living soul"—around. I was cold, my feet were cold, and I was running out of daylight. Suddenly I said out loud, "Aunt Minnie, I know you want this wreath, so you'd better help me or someone else gets it." At that instant, I turned, and right behind me was Aunt Minnie's bronze marker. There wasn't a speck of snow on it. The entire marker was exposed, while Uncle Billy's marker was covered with snow. This blew me away. I had tried so hard to find her grave and had been ready to give up. All I could say now was, "Aunt Minnie, I knew you wanted this wreath."

Gene, Daddy, and I left for St. Louis on the 21st. Daddy did well on the trip. He had the entire back seat, so he was able to sleep. We stopped at each rest stop and pretended we had to go to the bathroom. This was the only way we could get Daddy out of the car. When we'd ask, he'd always say, for some reason, that he didn't have to go to the bathroom. But when he did have to go, it was right then! So we had to persuade him to go and try. He slept most of the way, though. The trip seemed very tiring for him.

When we pulled up in front of Kim and Toren's house, I was so excited. The front door opened, and there stood Kim. She looked very tired. She was holding this tiny, beautiful baby. I couldn't get my coat off before I reached for my new grandson. I think I kissed Kim, but I'm not sure. I just remember tiny Brandon Luke, wrapped in his blanket, his big brown eyes looking up at his grandmother. I was ready to burst. Gene and Daddy were both trying to see as I uncovered him to show him off. We all thought he looked so little, but his birth weight—6 lbs. 12 ozs.—wasn't bad, especially since he was a month early. If Kim had gone full-term, he'd have been a whopper.

When Daddy was seated in the family room, Kim let him hold the baby. Daddy's face was as bright as their Christmas tree. He started on that day calling the baby "Amigo." We all just stood back and watched Daddy holding Brandon. No words were spoken among us, but I believe each of us—Toren, Kim, Gene, and myself—felt something powerful seeing Daddy smiling, cooing, and loving this Christmas Blessing from Above.

The following day was Christmas Eve. When Toren's family came in that evening, Baby Brandon was dressed in his red velvet Santa suit, hat and all. His big brown eyes looked around as if to say, "Who are all these folks? Where am I?" He was so cute. Yes, my grandson was so cute! There must have been thirty people there that evening. There was plenty of good food, laughter, joy, and love. Daddy, Gene, and I were able to join in the celebration of this day without dwelling on the painful experience of the previous Christmas.

Although Marsha wasn't with us, but home in Maryland with her family, we called back and forth, making her a part of what was going on in St. Louis. Neither Marsha, nor myself had forgotten Momma and Aunt Minnie. I chose to believe that this new, healthy baby boy had been sent to us from above. His arrival into our lives, at this particular time, allowed us to be lifted from despair. We were given a chance to focus on the gift of life and the abundant blessings we had received. Marsha and I both cherished this uplifting.

It wasn't just Marsha and I either. I could see happiness and joy in my father's eyes, too. He sat in the chair like a statue, holding Brandon. Even though he couldn't see him very well, Daddy felt the warmth of this new life. This was *his* blood now flowing in another generation. This gave him great pleasure each day. Often we would look over, and both Dad and Brandon were asleep. Kim and Toren were as pleased as Gene and I, watching Daddy holding Brandon. This was one trip when he wasn't worried about getting back home. What a Christmas Blessing!

On the day we left, Kim stood in the dining room window as we pulled out of the driveway. Toren was at work.

She later told me that, when we drove away, she went upstairs, put Brandon in his bassinet, and lay across their bed crying and crying. She was depressed and the exhaustion had caught up with her. She drifted off to sleep. When she awakened, the sadness was gone, and her beautiful son was hungry. She sat in her new rocking chair while Brandon nursed. And Kim sang to her son!

The Lord makes His Divine presence in our lives known to us in many, many ways. He is there to comfort and protect. He is there to rejoice and spread His love. We must recognize His goodness, even through pain and sorrow. It's not always easy. But reflection allows each of us the opportunity to fully understand His mercy and grace, in our lives. Thanks Be to Our Lord. May we never forsake Him.

Four Angels at your Bed,
Two at the Foot,
Two at the Head,
Four to Carry You,
When You Are Dead.
Good Night!

about the author

Carol Ann Walker Bouldin was born in Cincinnati, Ohio, in 1943, to Albert James and Harriet Ferguson Walker. A graduate of Ohio University, Carol also earned a Master's in Educational Administration from the University of Cincinnati. She was with Cincinnati Public Schools for thirty years, moving from teacher to principal to district-level administrator. After her retirement, she served as Assistant Director of the Children's Services Division of the Hamilton County Board of Mental Retardation and Developmental Disabilities. In 1996, Carol established Education Consultant Services, advocating for children with special needs.